HYDRAULIC DIE FORMING FOR JEWELERS AND METALSMITHS

Susan Kingsley

20-TON PRESS
Carmel, California

Hydraulic Die Forming for Jewelers and Metalsmiths

Third Edition

Library of Congress Catalog Card Number: 92-83947

20-TON PRESS
P.O. Box 222492
Carmel, California 93922

Printed in the United States of America
ISBN 0-9635832-0-4

Cover: "Amphora Ag 1," Susan Kingsley, 1990, 3 ½"x ½"x ¾", Sterling Silver. Photo by Lee Hocker

To my father, Edwin David
and to the memory of my mother, Marion Clay David,
who taught me the value of making things
for use, for pleasure and for giving.

contents

preface

Making things has been my primary interest for as long as I can remember. As a child, I sewed clothes, made beaded headbands, wove potholders, embroidered pillowcases and enameled copper jewelry. When I was 9, I decided to be an artist. Metalsmithing entered my life much later. I "found" metal long after studying fine arts in college, painting, teaching, moving, raising children and exploring various other mediums. Metalsmithing has allowed me to combine my interest in making things with making art. I especially love the technical challenges of realizing my aesthetic ideas in metal, but because traditional metalsmithing methods are not always suitable, I continually look for ways to individualize my work.

My interest in the hydraulic press began in 1978 while looking for an efficient way to make small hollow forms. An article by Marc Paisin in *Goldsmiths Journal* (Volume III, No. 6, December 1977) outlined the use of Devcon® Plastic Steel® for conforming dies. It was a process he devised from Masonite® die techniques developed by Richard Thomas and Val Link at Cranbrook Academy during the 1960s and on work done by Ruth Girard at the University of California at Berkeley in the 1970s. Robin Casady designed the adjustable press that I used and my introduction to the possibilities of using urethane for forming in the early 1980s should be credited to him as well.

I didn't take machine shop in high school and I have no formal background in science or engineering. What I know about die forming I have learned through experimentation, following up on "what ifs" and by accident. Because I had no preconceived notions or expectations, I used the problem-solving approach of an artist. I got results, great results, more results than I knew what to do with.

In keeping with the American tradition of sharing craft techniques, I wrote an article on die forming that was published in *Metalsmith* (Summer 1985). This eventually led to invitations to teach workshops which have provided many opportunities for me to learn from the questions and ideas put forward by wonderful, like-minded people from throughout the country.

Early in 1990 I began working with Lee Marshall, an engineer and inventor who had designed a small hydraulic press. He started a company, Bonny Doon Engineering, to manufacture his press and develop tools for die forming and metalsmithing. The availability of his standardized "ready-made" press, the "Bonny Doon," has opened this process to a larger audience and the addition of customized hydraulics, accessories and specially prepared urethanes has added to the safety and range of possibilities for die forming. I am grateful that Lee has taken an interest in the development and marketing of tools that fit the special needs of studio jewelers and metalsmiths at a time when many of our suppliers are abandoning us in favor of tools appropriate for mass production and the jewelry industry.

My primary interest is in the creative use of dies for jewelry and metalsmithing in a studio setting and in sharing information with others. Die forming has helped me to be more efficient and productive, freed me from many of the constraints inherent in other processes and continually suggests new areas for exploration. My purpose has never been to replace handwork, to duplicate processes used in manufacturing or simply to make things fast and cheap.

Perhaps I should admit that die forming is, to use a classic California expression of wonderment, a trip! For those of us who can't bench press 20 tons, die forming is a chance to play Wonder Woman or Superman. You have infinite power! (Well, 20 tons.) This "magic" hydraulic power makes die forming more a mental process than a physical one, and the playing field is level. Everybody has the same 20-ton strength and there is art in how you apply it.

Along with describing how to make and use different dies, I have tried to explain the principles which I hope will enable you to apply them to your own work. I have intentionally cited examples that are incomplete in the hope that the process will be used in individual ways. They are my own unless otherwise noted.

Many people provided invaluable help as I developed and wrote this book. I would like to thank Lee Marshall for answering my many technical questions, David Shelton for freely sharing his hard-earned knowledge of how to heat treat blanking dies, Monona Rossol for answering many questions about health issues and Marc Paisin, a man with many talents, for many kinds of advice. I am also grateful to Robin Casady for helping me recover after several Mac crashes, to Joanne Crosta for her skillful editing, to Lee Hocker for his expertise in photography (unless otherwise noted, all photos are Lee's), to Ed David for his careful reading and help with the index, to Gary Russell for his proficiency in the darkroom and, most of all, to Jess Knubis for his excellent book design and illustrations. Finally, I would like to thank my husband, Ed Kingsley, without whose patience, encouragement and support this book would not be, and my sons David and Brian for being who they are.

introduction

The basic properties of metal have not changed since ancient times, nor has human nature. There have always been metalsmiths who defended their old tools and methods, viewing change as a threat to tradition and their livelihoods, and innovators ready to explore better and easier ways to practice their craft. Even today, many people assume that the use of mechanical advantage, machines or new materials is a rejection of craft. Forming metal with dies, in particular, is often regarded as an industrial process, suitable only for the mindless manufacture of necessarily uniform, boring objects. Out of a desire to keep craft "pure," they dismiss not only a very useful process, but all objects made with dies.

This is an unfortunate misunderstanding of both die forming and craft. Dies have, in fact, been used for making jewelry since ancient times. Etruscan and Greek goldsmiths carved designs into bronze plates, and hammered thin gold sheet into them through thick sheets of lead. Necklaces, bracelets and earrings made from die-formed gold and silver flowers, leaves, acorns, human figures and animals were found at Pompeii. By the early 16th century, dies were employed in presses. In his treatise on goldsmithing, Benvenuto Cellini described how he used a screw press to make medals for Pope Clement VII. The hydraulic press has been around since 1796 when it was invented by the English mechanic Joseph Bramah.

22K Gold Pendant, Linda Threadgill, 1990, 3 ¼" x 2 x ½", photo by Jim Threadgill

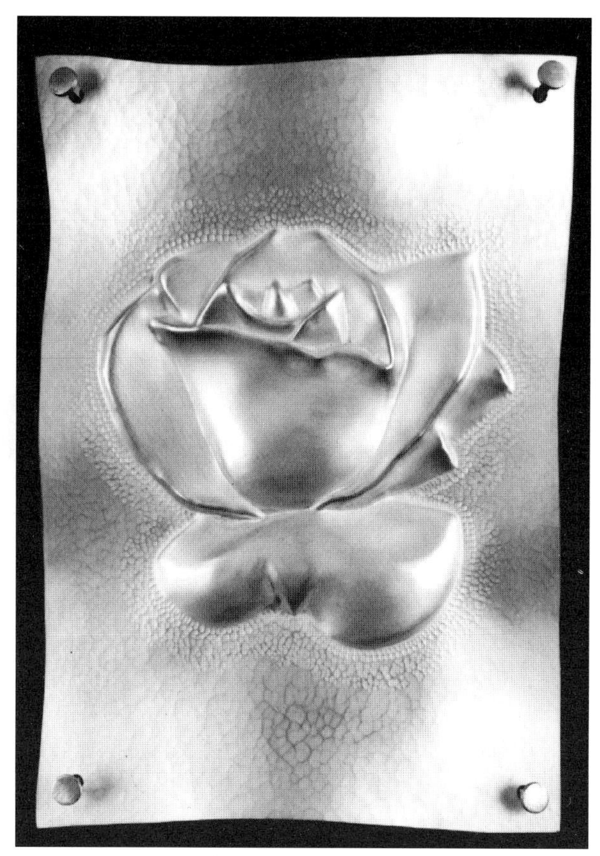

"Images of Perfection, Rose," Kate Wagle, 1991, Sterling, 6"x 4"x 3/8", photo by Richard Gehrke

In the search for improved jewelry-making methods, craft has traditionally made use of the materials at hand. Over the centuries, gold casting became a highly developed art form in West Africa where gold was abundant, and the art of working and carving wood was central to Native American peoples of the Northwest Coast. Thermoplastics, urethane and epoxy for die making are then very "natural" materials for use in our hi-tech 20th century global village. If they are not available at the corner hardware store, there is usually an 800 number to call.

Die forming metal with a hydraulic press and using modern materials to make dies follow the tradition of innovation in craft. Although there has always been resistance to change, I have no doubt that some Etruscans would have tossed their bronze dies if they had steel or that Cellini would have used a hydraulic press had one been available in 1525.

"While Walking–Morrissey Street," Lynda Watson-Abbott, 1991, Fine and Sterling Silver and 18K Gold Brooch, 2 1/8" x 2 5/8" x 3/8" photo by Lynda Watson-Abbot

This book is about how metalsmiths and jewelers can take advantage of the intrinsic characteristics of metal, understood since antiquity, and employ a mechanical device known for nearly 200 years to make objects by hand. Die forming, as described here, does not restrict or replace hand work. In fact, the die-forming process requires additional skill and knowledge in the making of dies, and in controlling their use in the press. Die forming extends the possibilities of what one can make and opens new avenues of exploration. It is a continuation of the tradition of innovation in craft.

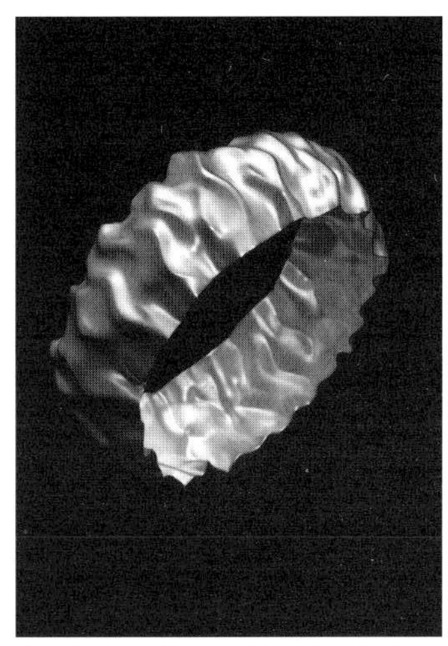

Bracelet, Susan Kingsley, 1990, 3" x 3" x 1" Sterling, photo by Lee Hocker

The processes that are described in this book encourage the use of metal as a responsive plastic medium. The ease with which forms can be produced leads one away from the tendency to treat sheet metal as if it were plywood. Whether metal is shaped with successive hammer blows or with a die, metal is metal. Die forming doesn't make skill unnecessary, hammers obsolete or understanding metals less important.

Die forming presents a number of advantages over traditional metal-forming techniques. For example, when doming metal into urethane, pressure is simultaneously applied over the whole surface of the metal. This is far less stressful to the metal than using a hammer or mallet. Mokumé and married metals can be formed without splitting. Roll printed or etched metal can be formed without damaging the surface and there are no hammer marks.

Because formed metal has more strength, die forming permits the use of thinner gauge material. Thus, jewelry can be made lighter in weight and more comfortable, with material cost reduced when working in precious metals.

Dies are tools that shape and mark your work in the same way that a particular hammer or stamp does. While they can be used to make reproductions, dies should not be considered "molds" capable only of the reproduction of identical objects. The manner in which a die is used will make variations and individualization an inherent part of the process.

Ruby Flask, Sue Amendolara, 1992, 6" x 3 ½" x 2" Sterling Silver, Avonite, Enamel, 24K Gold Foil, photo by Rick Potteiger

"Jack is Expensive," Christina Smith, 1989, Sterling Brooch, 4 ¼" x 3 ½" x ¾", photo by David De Vries

The processes described in this book are intended for nonferrous metals. Copper, sterling, fine silver, niobium, pewter and aluminum are easily formed with dies. Yellow golds and brass form well as does titanium. Stiffer metals such as nickel, nickel silver and bronze are not recommended.

Die forming is an environmentally "clean" process. The hydraulic press is self-contained, quiet, efficient and nonpolluting. In addition to forming metal, the press replaces the need for many separate tools in the studio. With accessories and adaptations, you can blank, stamp and bend metal. When used as suggested, the equipment is easily maintained and the process is safe.

Several of the materials listed here for making dies pose potential health hazards. If you choose to use them, safety procedures must be understood and followed. Please consult *The Artist's Complete Health and Safety Guide*, by Monona Rossol, to learn how to use potentially toxic materials safely and ethically and note the disclaimer at the front of this book.

I wanted to be consistent with one system of measurement, but it just wasn't practical. We still use inches and fractions for purchasing some

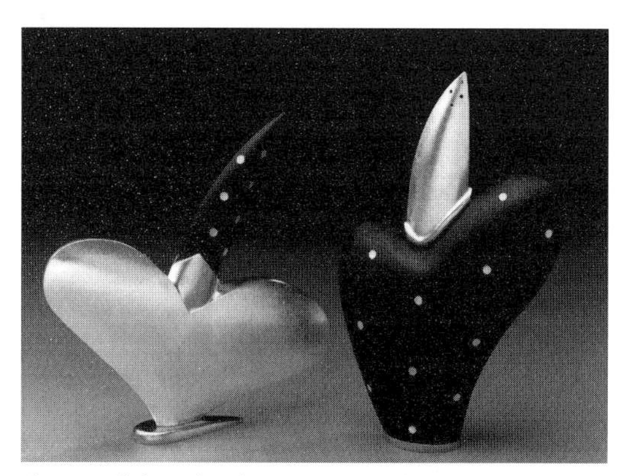

"Dancing Shakers Salt and Pepper," Sue Amendolara, 1992, Sterling Silver, Ebony, 24K Gold Foil, Mother of Pearl. (L) 4½" x 5" x 2", (R) 5" x 3" x 2", photo by Rick Potteiger.

materials and for visualizing size and relationships, millimeters for accurate small measurements, gauges for metal thicknesses, grams for weighing, etc. I have used whatever seemed the most common measurement for the particular purpose and in several instances include two. I have also provided tables in the appendix to make some of the calculations unnecessary.

Throughout the book, corresponding photo or illustration numbers appear directly before the accompanying text.

Hydraulic die forming is a metalsmithing process that empowers the artist with superhuman strength, increased productivity and the potential to create in ways that cannot be achieved with other methods. Because dies are both made and used by the artist, work is individualized. While a hydraulic press can be used to make standardized objects in quantity, die forming has a history and a potential for more creative use, and that is the subject of this book.

"Homage to Georgia O'Keeffe," Susan Kingsley, 1987, Sterling and copper brooch, 6 ½"x 7" x 1 ½", photo by Lee Hocker

chapter I
equipment

press frames

Many welded press frames built by metalsmiths in the late 1970s and early 1980s are still in use. The frames were constructed of "C" channel or "L" angle steel. Most have a fixed upper platen and a free middle platen that rests on the jack within a welded section of pipe. Many of these presses were built with materials at hand, primarily for use with conforming dies. There is a lot of variation in their size, design and condition after years of use (and misuse).

Small presses that were built for other purposes, such as making rubber molds, turn up occasionally at tool sales or flea markets and may be usable. Some of these presses incorporate heating elements and include built-in power sources. Inexpensive automotive shop presses available through discount tool suppliers are not recommended. They are unnecessarily large and often underpowered. In addition, they require numerous modifications, seem to bend and break easily and cannot be used with the bolt-on accessories.

In order to be suitable for all the applications described in this book, press frames should meet the following standards: 1) the platens must be flat and at least 1-inch thick (optimum platen size is 6x6 inches), 2) the frame must be square and the platens perfectly parallel, 3) there must be at least six inches clearance between the platens, 4) the upper platen must be drilled to accept accessories, and 5) the frame should be securely fastened to a bench. Return springs are optional, but make the press much more serviceable.

1

1 A simple design for a "homemade" press is the bolted together **adjustable press** designed by Robin Casady. The stationary top and bottom platens are connected by threaded rods separated by tubing spacers. The threaded rods run down through holes in the bench and are secured by nuts.

The distance between the platens can be adjusted by adding or removing sections of tubing. The middle platen (which is drilled with larger holes) rests on the jack and moves up and down on the spacers.

The platens must be level and parallel. Grease the tubing in the areas where the middle platen slides.

Plans for building the adjustable press are included in the appendix. The correct thickness for the steel platens, the size of the threaded rod and the number of nuts needed to withstand 20 tons were calculated by a machinist. Most machine shops or metal-fabrication businesses are able to supply the steel parts you need: threaded steel rod, steel tubing, nuts, washers and the steel platens. They will also cut and drill. It is important that the holes are drilled accurately and that the spacers are cut precisely. The springs are found at most hardware stores.

This simple design has several advantages. Its height (and clearance space) is adjustable, it can be built without welding and it may be used with a variety of different hydraulic systems. The cost of building this press depends on your source of supply and how much of the work you do yourself. Due to its weight and labor costs, it could be quite expensive. You should get firm quotes on costs before you begin.

2 The **Bonny Doon press** was created for the applications described in this book and satisfies all five requirements listed on page 16. The frame is welded from heavy-gauge, 2-inch-square steel tubing. The 6-inch square steel platens have 6 inches of clearance. The upper platen is drilled to accept accessories for bending and forming. Substantial springs inside the uprights hold the lower platen in position and keep it level. The press is engineered for use with Bonny Doon's own, specially modified hydraulic systems. However, the frame may adapted for use with other hydraulic systems, such as those as described in the following pages. Consult the appendix for details.

2

hydraulic power

Ten-ton jacks are adequate for some processes, but up to 20 tons of power is necessary to perform all the processes described in this book. Hydraulic power can be applied progressively — the reason it is so versatile and has so many uses. Very little pressure is used for some processes, such as blanking and forming in conforming dies, while others, such as embossing, often require the full 20 tons.

When using blanking or small conforming dies, power is applied lightly at first, and then increased slowly. You know when to stop because you can hear or see results. But in other processes, such as forming with urethane in matrix dies, you can't see what is happening to the metal. Although resistance increases as you pump the jack, there is no way to judge how much pressure is being applied or to know how much is needed. When you hear something "pop," it usually means the metal has split and you should have stopped applying pressure sooner.

When forming metal using hand processes, you learn to sense when the metal is work hardened and will stretch no further. The hammer makes a different sound and you no longer see the metal changing shape. You learn to stop and anneal before the metal becomes brittle enough to crack or break. This point is reached quickly in the press and is often difficult to predict.

Consistent results with some die-forming processes are achieved only if the jack has a gauge. The gauge doesn't tell you when the metal becomes work hardened or how much force is actually being used. What is measured by the gauge are the psi, pounds per square inch, being exerted by the hydraulic ram. There can be no way to predict how many psi will be required because each die is unique in size and shape and different metals and metal thickness also affect the outcome.

Consistency is possible because a specific "reading" will always produce the same result. You must experiment to find what works for each die, recording the pressure readings, the metal gauge and number of annealings. Then you won't have to experiment the next time you use the die. This isn't difficult and only applies to some die-forming processes.

A table in the appendix (p. 75) lists the average ram force needed for various types of dies and compares the readings of different systems. Although this gives a rough idea of the readings that might be expected, it really depends upon the type, size and complexity of the die and the type and gauge of the metal being formed.

Bottle jacks, as they are sometimes called, are available without gauges at automotive and industrial supply houses. It is possible to modify a bottle jack with a pressure gauge, but the undertaking is technically difficult and potentially dangerous. Jacks vary internally. You must drill and tap the correct hydraulic passage or you will ruin the jack and may injure yourself. Because of these factors, I recommend purchasing this option from one of the suppliers listed in the appendix.

An alternative to the bottle jack is a **hydraulic system made up of components,** such as the one pictured with the adjustable press on page 16. It consists of a separate two-speed hand pump and a cylinder connected with a hydraulic hose and gauge. This type of system was assembled from components made by Enerpac®. The advantages are that the hand pump works more easily, the gauge is large and easy to read and local servicing is widely available. The disadvantage is that this system is considerably more expensive.

3 And finally, there is the option of a **hydraulic jack operated by an electric pump.** Instead of hand pumping, you simply press a button. Because of its speed and ease of use, this system is recom-

3

mended for production work and for anyone requiring a process that is less physically demanding. While speed and power are good things, you loose the level of control provided by a hand pump. In order to achieve accurate and consistent results, and pinpoint control that is superior to the hand pump, I recommend adding a pressure regulator and using a large gauge. This system is manufactured by Power Team.

Keep the following in mind when using a hydraulic jack:

• *Never exceed the maximum rating for a jack.* For a 20-ton jack this is slightly less than 9000 psi. If you don't have a gauge, don't continue beyond the point when the jack handle begins to bend. Caution: adding a longer and/or steel pipe handle to the jack makes it easier to pump, but all too easy to go beyond what is recommended.

• *Hydraulic jacks work most efficiently when the ram is not fully extended.* In other words, if the ram is six inches long, use one or more spacers so that when you begin feeling resistance or registering pressure on the gauge, the ram is extended no more than four or five inches.

• It is a good idea to *leave the ram in the "down" position when the jack is not being used* for any length of time.

• Although the jack does not normally consume oil, you may occasionally need to add some. Use only hydraulic jack oil. It can be purchased at a hardware or auto parts store. To add oil, the jack must be removed from the press. Tilt the jack on its side and remove the stopper from the fill hole with a pair of needle nose pliers. You can use a funnel but it is much faster if you use a "trough." (Modify a small funnel by making a cut from top to bottom.) Fill to the level of the hole and re-insert the stopper. An $\frac{1}{8}$-inch "drift punch," available at automotive stores, is handy for this purpose.

• When jacks receive continual use with the ram in its extended position, two problems may occur –

at mid-range the jack will not work at full stroke, and when the ram is left in the down position, oil will leak from the control valve.

4 *Always use one or more spacers with the urethane when using matrix or embossing dies so that the ram is not fully extended during use.*

To prevent or fix these occurrences, you can make a small hole in the rubber "stopper" over the fill hole. Lay the jack on its side, take the rubber stopper out of the fill hole with needle nose pliers and drill a small (i.e., #60 drill bit) hole in it. Replace the stopper with a drift punch.

urethane

Urethane is a tough rubber-like substance. Used in conjunction with non-conforming dies, it becomes under pressure the other half of the die. 4 Unlike rubber, which compresses, urethane "flows" under pressure, distributing its force evenly over a large area. After use, urethane returns to its original shape and can be reused thousands of times. Even its ragged appearance after extended use does not affect performance. Ordinary rubber can be used, but in comparison to urethane, the results are unsatisfactory. Rubber simply does not form metal as well and it degenerates quickly, making results unpredictable.

Urethane forms metal gently and evenly, leaving a smooth, unmarked surface. Finishing time is greatly reduced. Some distortion occurs when a plane becomes contoured, but metal that has been etched, roll printed, engraved, textured, reticulated or slumped can be formed without damaging the surface. And because uniform pressure is maintained over the whole surface of the work, mokumé, married metals, soldered inlay and laminated work are formed with less risk of splitting. Urethane is manufactured in different degrees of hardness measured in durometers: *95 durometer is the hardest and used most often; 80 durometer is slightly softer, moves under less pressure and has specific uses.*

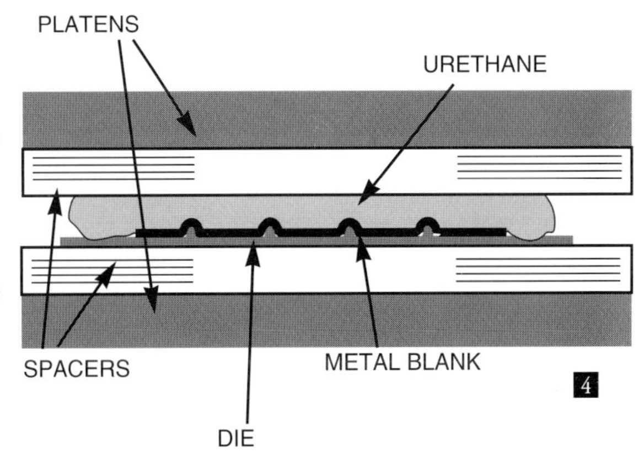

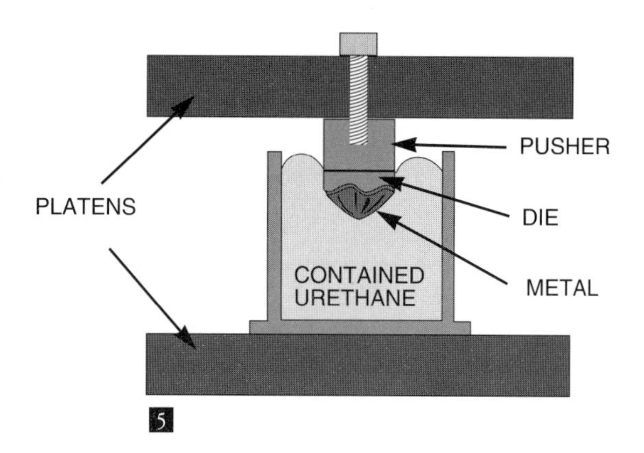

5

6

7

5 *Because urethane "flows" under pressure, contain it whenever possible and force it to flow around or into a die, rather than out and away.* 6 Contained urethane blocks consist of sections of ¼-inch walled steel pipe not quite filled with urethane. Bottomless containers may be used, but those with welded bases function even more efficiently.

7 **Urethane pads** are made in various thicknesses and used with many of the forming accessories and dies that do not fit into the contained block. The thickness used depends on the depth and size of the tool or die. Larger dies, in which deeper forming is desired, require thicker pads (½ to 1 inch). Very thin (⅛ and 1/16 inch) pads are used for embossing. Urethane pads should be used with one or more spacers in order to avoid "working" the ram at the top of its range. Acrylic blocks, 6x6x1 inch, make ideal spacers (see illustration on the previous page). Avoid using spacers that are smaller than the die or made of compressible materials (such as wood or Masonite®).

Urethane pads and containers are supplied by Bonny Doon in the sizes and durometers which have proven most useful to metalsmiths. I find it useful to have a variety of urethane pads and blocks available in different sizes, thicknesses and durometers. In order to distinguish between the durometers, 80 is yellow or yellow-orange, and 95 is orange or red. Large sheets of urethane may be available from industrial suppliers in various colors.

When I first began using urethane for die forming, I used a two-component material called Flexane® to make pads and contained blocks. I stopped doing so after learning that dangerous toxins are released during mixing, pouring and curing. Air-purifying respirators do not provide adequate protection and even very low levels of exposure may cause severe reactions. In addition, *ACTS FACTS* (September 1990) reports that chemicals found in Flexane® have been identified as possible carcinogens. (For information on a wide variety of materials and processes used by artists, see *The Artist's Complete Health and Safety Guide*, by Monona Rossol.)

If you already have Flexane® pads and blocks, they are safe to use. To cut urethane (including Flexane®) use scissors or a sharp knife. Never heat, burn, saw, sand or grind urethane as toxic gases are again released and even small amounts may be harmful.

8

accessories

8 The press may be used for doming, forming, bending and stamping using accessories that have been developed by Lee Marshall. Any press with the three-hole pattern in the upper platen and a 6-inch working space can be used with these accessories. (See appendix p. 76 for hole pattern.)

9 **"Pushers"** are 1-inch deep steel cylinders that bolt to the top platen of the press. Because urethane does not fill the containers, pushers are needed to extend and press dies down into it. Pushers also function as spacers by making the working area of the punch more than an inch from the top of the ram where the jack is more efficient.

Most of the accessories function as non-conforming dies. They are designed to be fastened to the top platen of the press and used with urethane to form and bend metal.

9

10

11

12

13

They can be used as "generic" tools, as is, or modified and adapted to meet specific needs. 10 The **master holder** holds tool shanks in a vertical position in the center of the top platen. It has a brass-tipped set screw and will hold shanks ¾ inch to 1 inch in diameter.

11 The **T-shaped forming tools** are designed for use in the master holder and may be modified by grinding into different shapes. They are usually used with 1-inch-thick 90 durometer pads.

Always employ annealed metal and avoid pressing metal with sharp edges into the urethane. *Watch what is happening in the press rather than a pressure gauge.* Press until the metal is buried in the urethane. How quickly the metal forms depends on the overall size, hardness and thickness of the metal as well as the complexity of the form. It is not unusual for forming to be done in stages, with one or more annealings, rather than all at once.

12 13 **Large daps and mushroom formers** with 1-inch shanks are also used with the master holder. The daps form metal most

efficiently into 2-inch contained urethane and the mushrooms into 3-inch contained urethane. **14** Mushrooms are available with domes designated by the diameters of various spheres from 3 to 8 inches. The 3-inch mushroom forms the deepest dome and the 8 inch forms the gentlest. **15** **Steel cups** in radii matching the mushrooms are also available, offering the possibility of forming and embossing in one operation. (See illustration on page 68.)

16 Because the mushrooms are not hardened steel, they should be protected when used with the steel cups in metal-to-metal situations. Simply form a piece of 16-gauge brass over the mushroom and secure it with double-sided tape. The brass can be replaced when needed. **17** Tools with smaller shanks (from ⅛- to ¾-inch), including regular **dapping punches**, may be used in the master holder with a **collet** from the **dap adapter set.** Punches must be pressed into urethane carefully. When using smaller punches there is a danger of tearing the urethane or causing the shank to break under excessive force.

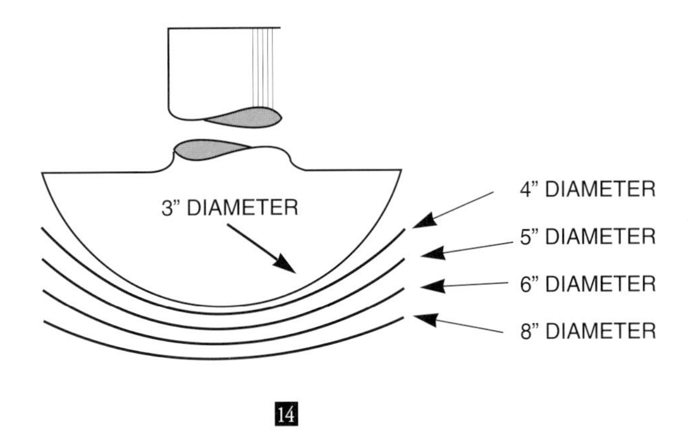

14

15

16

17

18

19

20

Standard ring stamps, hallmarking stamps and decorative stamps can be used in a hydraulic press. These stamps are also used in the dap adapter collets in the master holder. Stamps and hallmarks are used without urethane against a rigid surface. The advantage is that perfectly vertical alignment is assured and correct pressure can be consistently applied. Keep in mind, however, that *only a very small amount of pressure is needed and excessive pressure is very dangerous.* The steel shaft could punch through the metal being stamped and damage the face of the stamp, or even snap and injure the operator of the press. It is a good idea to do this type of stamping inside a section of 4-inch plexiglass cylinder. Then, if the tool should break, the parts cannot cause injury.

Bending dies are flat metal dies with an edge used to bend metal in various ways. 18 The **brake assembly** bolts to the upper platen and holds bending dies. It comes with a long **brake** for making 90-degree bends up to 6 inches long and a set of short **finger brakes** for making 90-degree bends from ½ inch to 5 ¾ inches in length. Finger brakes are used for making boxes and frames when it is necessary for the die to reach into a partially formed piece.

19 When using a bending die in the press, simply draw a line with a marking pen where the fold is to be made, place the metal on a block of urethane, align it with the edge of the brake and apply pressure until the angle reaches 90 degrees. The edge of the brake does not seem sharp, but it will cut the urethane if it extends beyond the metal you are bending.

20 You can also make your own bending or "brake type" dies from steel or aluminum. The bending edge is cut and filed into various angles and curves out of material up to ⅜-inch thick. 21 Successive bends made with brake type dies result in unique effects. 22 The brake assembly comes with a "**standoff kit.**" A long bolt and tubing spacer

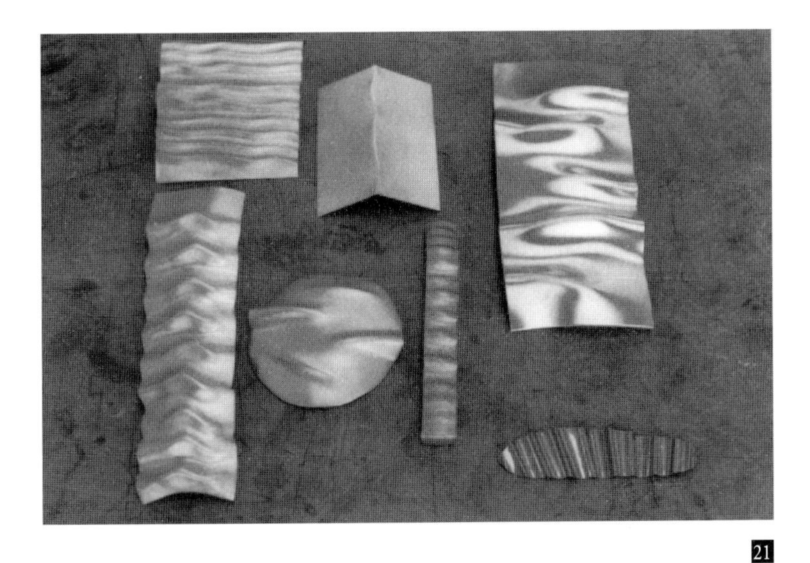

21

are used in the center to secure and distance the assembly from the upper platen. Solid spacers are placed on either end to maintain the distance. Dies should be used with an "outrigger" to minimize stress. 23 The stand-off kit makes it possible to form from the inside of a loop, such as a bracelet.

24 Another bending option is the **tubing former**. Steel mandrels from ¼ to 1 inch in diameter are used to form U-shaped channels or seamed tubing in 6-inch lengths. The steel mandrels are held on a "positioner" with rubber bands or double-sided tape.

25 To make tubing around a mandrel, it is necessary to calculate the *exact* width for the blank. The width depends upon both the diameter of the mandrel and the thickness of the metal to be used. A formula and table

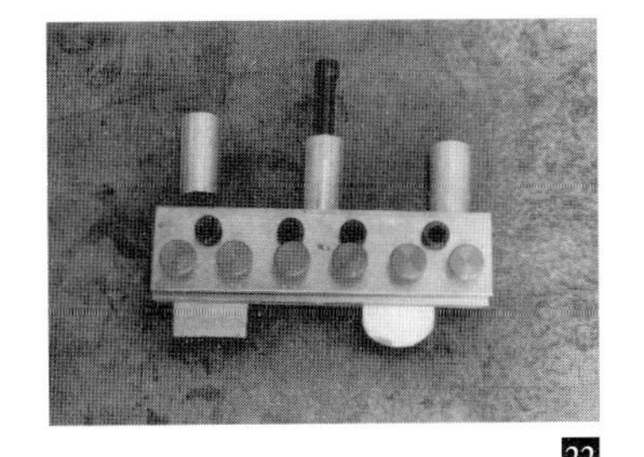

22

23

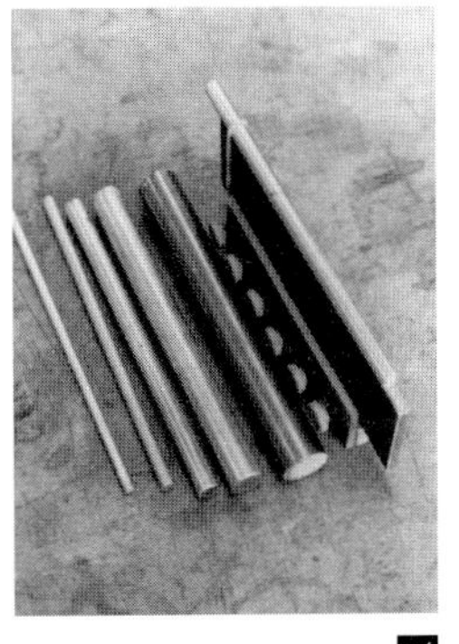

24

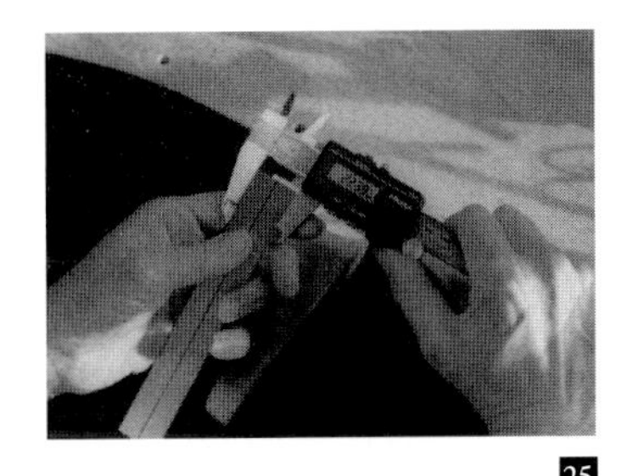

25

26

27

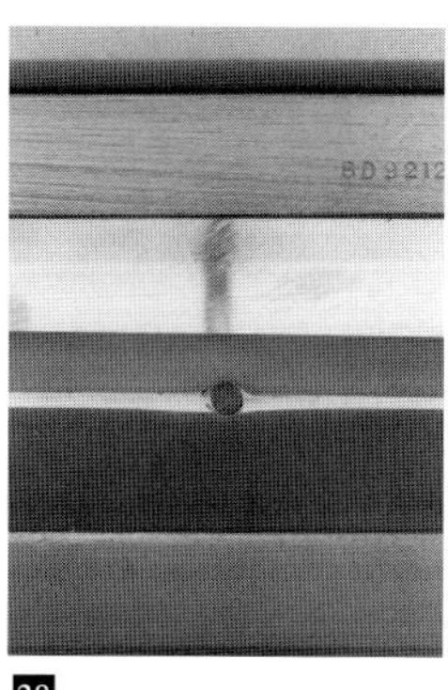

28

for determining blank widths is included in the appendix. After making the blank, draw a guide line down its center.

26 To minimize the finished seam it is necessary to bevel the edges slightly. This is easiest to do while the blank is still flat. 27 Begin forming with a mandrel larger than the finished tubing. Align the center mark under the mandrel, and press into a ½-inch or 1-inch 90 durometer pad until you have a U. After annealing the U, put it on the correct sized mandrel to close it. 28 Line up the center line of the U-shaped blank with the mandrel, and place it between two ½-inch pads with the opening to the side, as shown. (To maintain this alignment, you may need to tape the blank to the mandrel at each end.) Use a 95 pad on the bottom and 80 pad on the top, pressing down one side and then turning to press down the other side.

Finally, press the tubing and mandrel between two ½-inch 80 pads, again with the seam on the side. This will nearly close the seam. Remove the mandrel and anneal once more. 29 Then, without placing the tube back on the mandrel, gently close the seam with a leather or plastic mallet. It doesn't matter at this point if the tube is slightly out-of-round. 30 After soldering and cleaning up excess solder, reinsert the mandrel into the tube and planish the seam and tube. A little oil makes it easier to remove.

29

30

Tubes longer than 6 inches can be made in the same way, using longer mandrels. It works better if, in the first step, the mandrel is not fastened to the holder but slides along it. Begin forming in the middle of the blank and shape the metal in stages, a little at a time, working from the middle toward the ends of the blank.

A hydraulic press with 20 tons of power should not be considered a substitute for a kick press, but small (less than an inch in diameter) steel dies can be used for stamping (metal to metal) under certain conditions. 31 Two-part dies (called die sets) require the use of a "**die shoe**" for maintaining precise alignment. 32 Dies such as those illustrated must be professionally made. A supplier is listed in the appendix.

33 Standard **disc cutters** which are available from most jewelry suppliers may be used in the press. Circles up to an inch in diameter can be blanked easily from metals as heavy as 16 gauge. Care must be taken to avoid breaking the punch when using this type of die in the press. Once you hear the "pop" of metal being sheared, stop pressing! Remove the disc cutter from the press to recover the blank and separate the punch from the die.

"C" frame punches designed for industrial use may also be used in the press. Disc dies larger than one inch in diameter as well as other shapes may be obtained for use in the frames. A supplier for this type of unit is listed in the appendix.

31

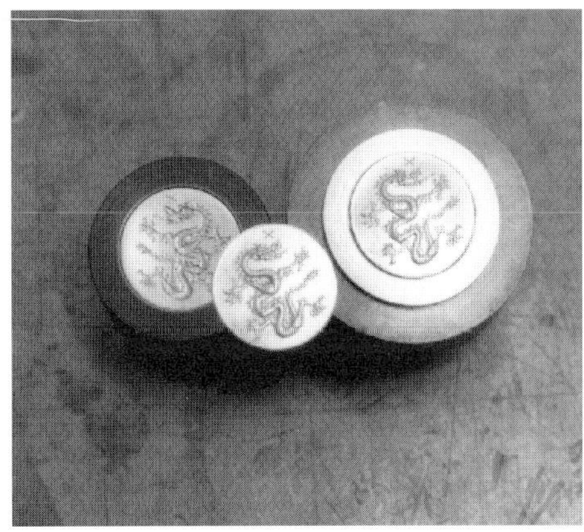

Die by Quicksilver, Michael Stewart design

32

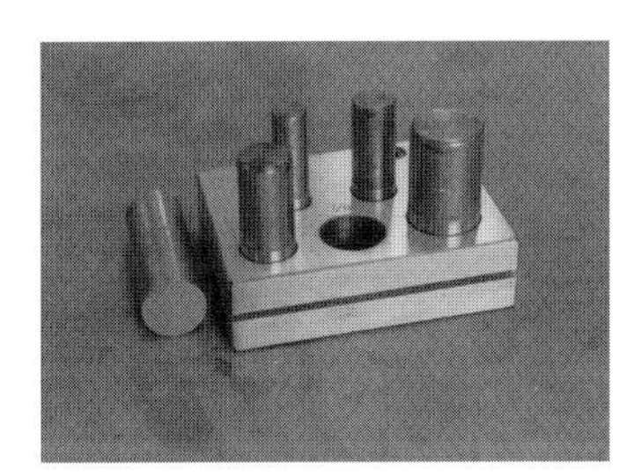

33

chapter 11
non-conforming dies

Non-conforming dies offer an opportunity for spontaneity and improvisation. They consist of only one part and can be made quickly from materials that are easily worked. Non-conforming dies are used in conjunction with urethane which serves as the other part. New designs can be developed, tested and re-worked without delay and expense and dies may be used in a variety of ways, resulting in different end products.

A number of inexpensive materials may be used to carve, construct or cast non-conforming dies. Probably the most useful material is cast thermoplastic acrylic resin, commonly called acrylic or plexiglass. It is available from plastic distributors under the brand names Plexiglas®, Acrylite® and Lucite®. It is important to obtain cast acrylic, rather than extruded glazing material available at hardware stores which is too soft. (I find Lucite® less satisfactory than other brands because it gets gummy when you saw or machine it.) Delrin®, another material available from plastic suppliers, is stronger and slightly more wear resistant than cast acrylic but is more difficult to work. Other materials useful for making dies include end grain hardwood, nickel, brass and steel. Aluminum is also useful in applications when wear is not critical.

1

punches

A punch is a tool that is driven against and into the metal to be formed, usually a "positive" of the desired shape. Urethane, when used with a punch, becomes the matrix or female part of the die. A dapping punch is a generic punch. A punch can also be called a "male" die.

1 A good punch that will hold up to hundreds of pressings can be made from a section of Delrin® or cast acrylic rod. A 1-inch section of an appropriate diameter is convenient and economical. It is important that the rod be cut accurately at 90 degrees with parallel faces. The Bonny Doon pre-domed punch blanks are a great time saver.

Because of the difficulty in accurately cutting this material yourself, I recommend having it cut by the supplier.

2 Acrylic can be shaped with coarse files or carved with burrs in a flexible shaft. Wear a dust mask and goggles when working with acrylic, and avoid grinding at high speed. Decomposition products are released during heating and are hazardous. If the acrylic becomes "gummy" or emits a sharp plastic odor, slow down and apply less pressure. Clean up the dust as soon as you are finished. Delrin® may be worked in the same way, following the same precautions.

To carve a punch, first rough in the shape with a coarse file or burr. *Punches are forming tools and contour should be the first consideration.* Make some pressings of the shape as you carve, before becoming too involved with detail and surface finishing. One advantage of this process is that you can test the die as you are making it. Avoid deep cuts and undercuts. Texture and detail can be part of the design but will be effective only with thinner metals. Formed metal has greater strength permitting the use of thinner, lighter weight metal, making jewelry more comfortable and less expensive.

3 Punches should be used with contained urethane. Use the container that most closely fits the size of the punch. The 95 durometer and the 80 durometer will give slightly different results. Bolt a pusher to the center of the upper platen of the press and use double-sided Scotch® tape to stick the punch to it.

4 The metal blank should be annealed and slightly smaller than the die. If a larger piece of metal is used, it will wrap itself around the die (like a bottle cap) and may be difficult to remove. It could also prevent the metal from being drawn into the die, causing a poor impression.

As you are pressing, watch the container rather than the gauge. The punch will go down into the container, burying itself and the

2

3

4

5

6

blank in the urethane. When using a container without a bottom, stop when the urethane starts to lift the container and push its way out of the bottom. Additional pressure is unproductive. If the container has a bottom, you can push the punch as much as two thirds of the way down. Avoid forcing the punch into the bottom of the container as this could damage or break the die.

Experiment with your dies, keeping a record of the results. 5 By using different metals, gauges and pressures, changing the shape and size of the blank and limiting how far into the urethane it is pressed you can achieve diverse results. Keep track of what you did so that you can repeat it. Write on each test the thickness of the metal used, the type of container, the number of annealings and any other variable.

6 *Keep in mind that the thickness of the metal is what determines how much detail will be achieved: the thinner the metal, the more detail.* These pressings illustrate (clockwise from left) the differences between 22, 24, 26, 28 and 30 gauge copper. To maximize detail, anneal and do repeated pressings. Partially formed blanks can be held in place on the die with a little masking tape.

matrix dies

7 A matrix die is simply a block of material with the outline of a shape removed, leaving its silhouette. It is the impression part of a die within which a form is shaped. Urethane, when used with a matrix die, becomes the punch.

Matrix dies are similar in concept to Masonite® dies, but it is not necessary to secure the metal to the die or to use a hammer to dap or form the metal. The urethane "clamps" the metal against the die and, under increasing pressure, pushes and stretches the metal into the silhouette. The result is a smooth, gently pillowed form with a flat flange and concise outline. There are, of course, no hammer marks.

Masonite® and aluminum can be used to make matrix dies, but I prefer cast acrylic sheet as it is inexpensive and easy to cut. Acrylic has a compressive strength of 18,000 psi and, properly made, will withstand considerable use.

8 The addition of a brass or steel face plate will prevent the edges from becoming rounded and is recommended if the die will be used for making hollow objects. It is possible to buy acrylic scrap pieces by the pound from plastics suppliers or to have larger sheets cut into usable sizes.

To determine the thickness for the die, measure the widest part of the silhouette. The wider the form, the deeper the relief you can press; the deeper the relief, the thicker the die must be. The accompanying chart gives a general idea of the sizes involved.

As a general rule the cutout part should be centered and no closer than ¾ inch to the edge of the die. A wider margin is even better because in addition to making the die stronger, it seems to make it more efficient. As the die is pressed into the uncontained urethane pad, the material must either squeeze into the opening, forming the metal, or flow out around the die. The wider margin seems to concentrate more

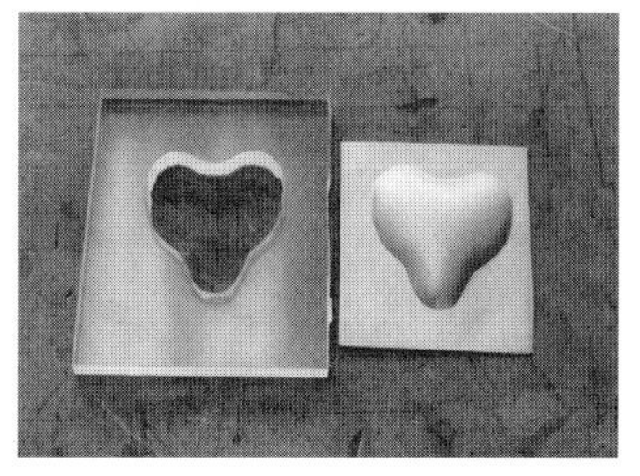

7

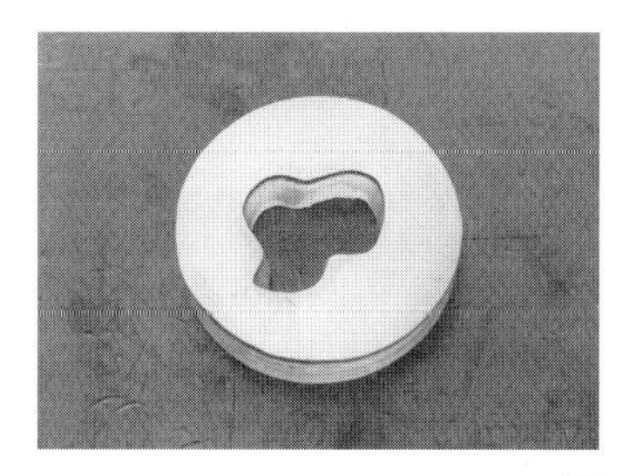

8

Width of silhouette	Thickness of die
1"	¼"
1½"	⅜"
2"	½"
2½"	¾"

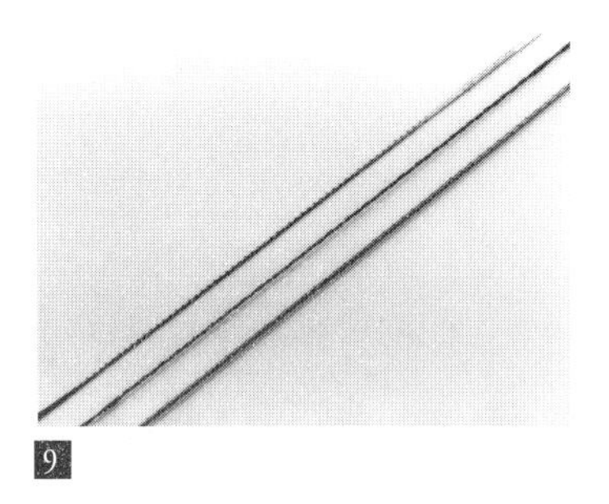

9

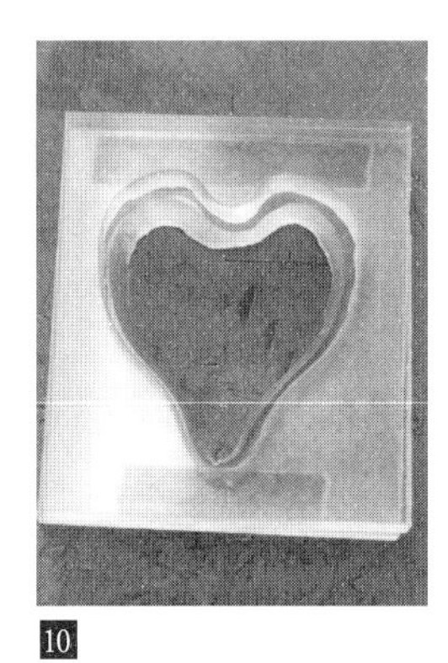

10

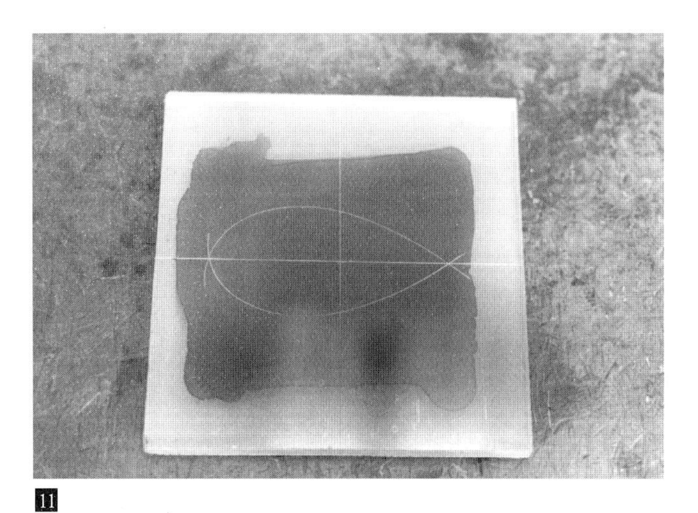

11

flow into the die and allow less to escape to the outside. Matrix dies that are in a round format, such as the one illustrated (8), can be used with contained urethane and are most efficient. (When using a round die with contained urethane, remember to use it with a pusher that is the same diameter or greater. The die must be fully supported.)

9 Acrylic sheet is most easily cut with a spiral, skip-tooth or mono-tooth saw blade inserted in a jewelers saw. Although these blades cut fast, they are not as smooth and accurate as a regular saw blade. Make your cut inside the line and then carefully file out to it. The cut should be vertical but its quality and finish are unimportant. The edge of the silhouette should be smooth, not rounded. This *edge* is the important part of the die.

10 Because of the difficulty of hand sawing ½-inch or thicker acrylic (as well as its greater cost and scarcity) you may prefer to build up the thickness in ¼-inch layers. *Make sure that there are no undercuts in the added lower layer.* You could permanently glue the layers together with acrylic cement, but double-sided tape works equally well. Some designs require greater thickness than listed in the chart. If your pressings

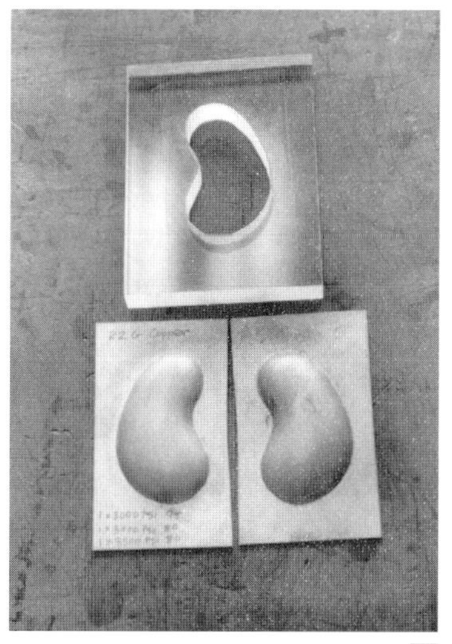

12

"bottom out" (press deeply enough to reach the bottom of the die, leaving a flat spot), you can simply add another layer to the die.

A matrix die with a symmetrical outline is a reversible matrix die because it can be used to form both the front and back of a hollow object. *Reversible dies must be made with the greatest accuracy if you want the edges to meet.* **11** To lay out a reversible die, remove the paper from the acrylic, coat the surface with layout fluid and use a scribe to draw the pattern directly onto it. (A penciled line on the protective paper of the acrylic is not sufficiently accurate.) Saw and file as described above.

12 An asymmetrical die can be used in turn from either side to make matching mirror-image halves of a hollow form if the cut is made at 90 degrees and the angle is perfectly maintained. It is not possible to saw free hand through a ¼-inch thickness with sufficient accuracy. **13** Reversible asymmetrical matrix dies can be made either with the Bonny Doon precision saw guide set at 90 degrees, as shown, or with a router, scroll saw or other device that ensures a perfectly vertical cut.

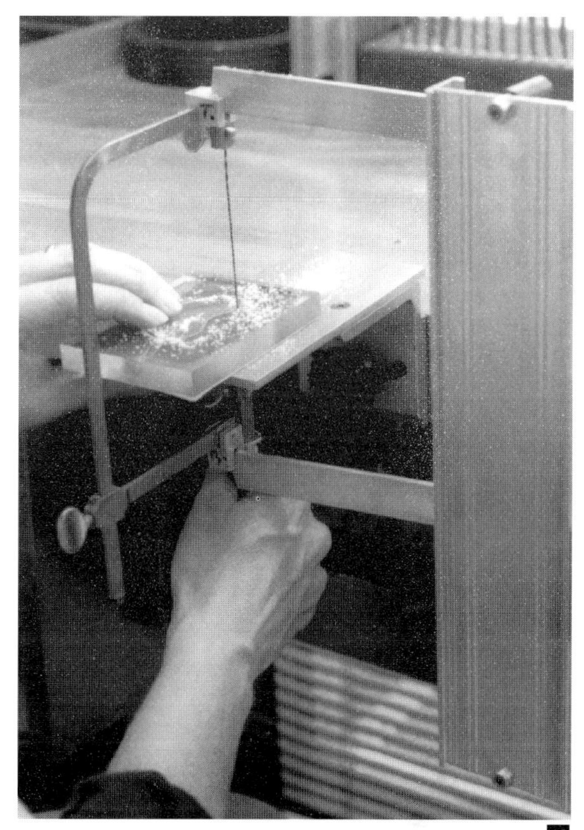

13

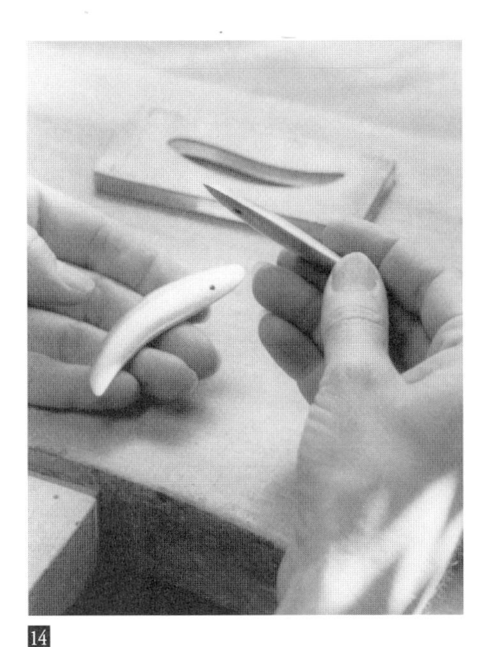

14

16

Another way to make an asymmetrical reversible matrix die is to make matching face plates. 14 This method is accurate enough to make hollow forms, and requires no special tools. You begin by gluing or taping together two pieces of metal the same size as your die. You can use either 16-gauge brass or $\frac{1}{16}$-inch steel. 15 After scribing the outline and drilling a starter hole, saw it out as if it were one piece. 16 Then file so that the edges are at 90 degrees. 17 When separated, the facing surfaces of the two plates can be used as face plates for a single two-sided acrylic die or made into a pair of dies. 18 Whichever you do, be sure that the acrylic is cut and filed to fully support the metal and that there are no undercuts. 19 The metal faces could either be glued in place or held with double-sided Scotch® tape.

Metal from 18 to 28 gauge can be formed in matrix dies. The thickness needed depends upon

17

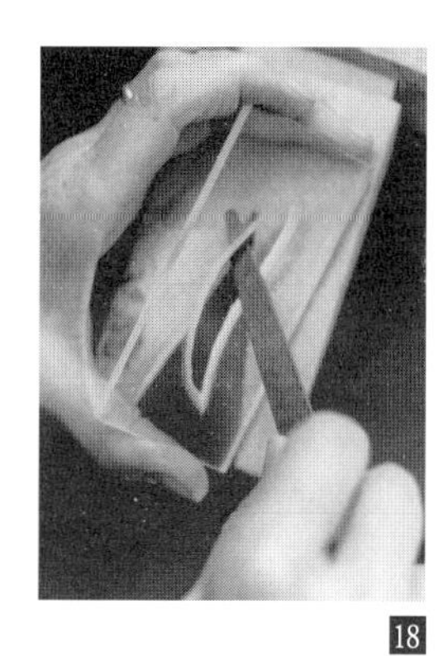

18

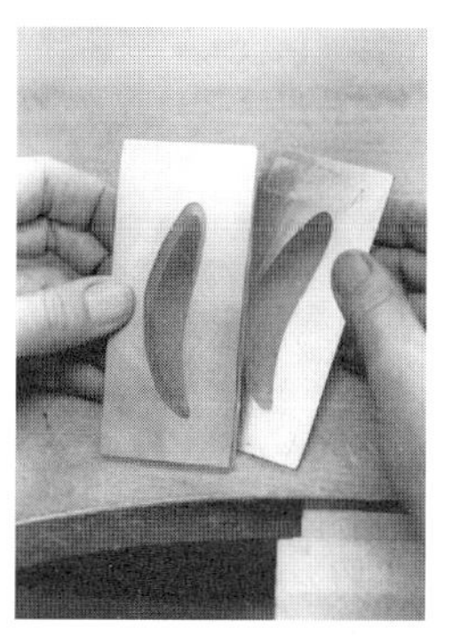

19

how much the metal will be stretched. Larger, thicker dies require heavier gauge metal. The smaller, thinner dies may be used to form thin metal. The way in which the formed metal will be used should also be considered in selecting metal thickness.

If the flange is to be removed, thicker metal should be used because the form will not be as strong. If a hollow object is to be constructed of two formed pieces, the metal should be thick enough to give you an "edge" to solder. The tendency is to use metal that is too thin. Keep in mind that the object you make needs to be strong enough to hold up to normal use.

Metal to be formed in matrix dies should be at least ½ inch larger than the silhouette because the flange must be between ¼ inch and ⅜ inch all the way around. The metal is not drawn into the die, but you need enough flange for the urethane to clamp.

Rectangular or square matrix dies are generally used with uncontained, 95 durometer (orange) pads. Use the 1-inch pads for larger and deeper dies (½ inch or more in thickness) and the thinner pads (¼ and ½ inch) for the smaller thinner dies (¼ to ⅜ inch).

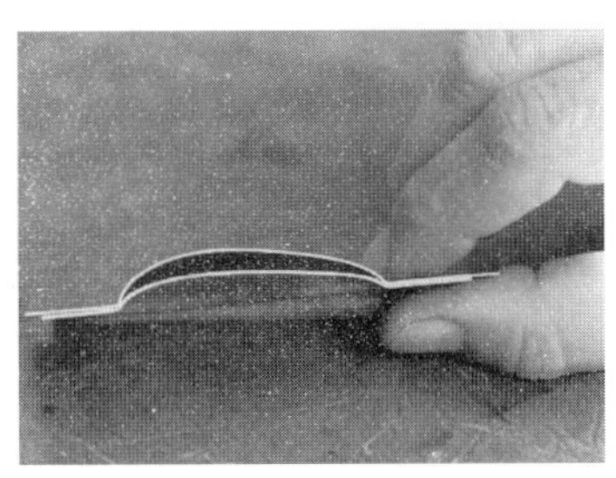

20

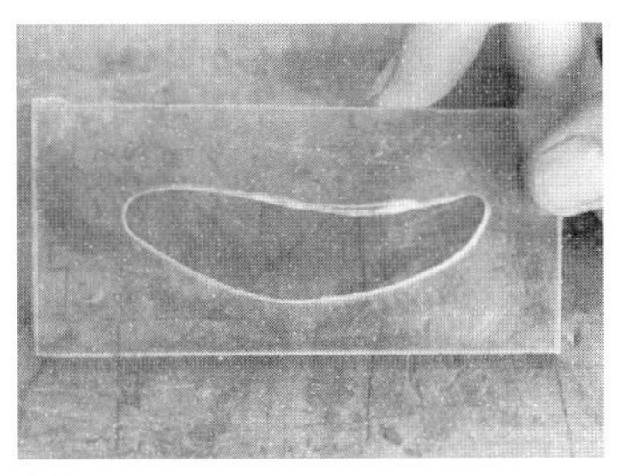

21

The 80 durometer (yellow) pads may also be used with matrix dies. The effect is a deeper, fuller shape that is slightly different than the 95. 20 A cross section of two pressings (made with an equal amount of force and the same gauge annealed metal) illustrates the difference between 80 and 90 durometer urethane. The 80 is "softer" and flows more easily, forming the metal deeper and faster.

This sounds like a good idea, but there are some drawbacks. The 80 draws metal into the die sooner than the 95, before it can securely clamp the flange against the die. 21 This drawing action can damage the edges of your die. And because the metal forms quickly with the 80, it also tends to work harden quickly. You have to be very careful or you will have torn or split metal.

In order to avoid these problems with the 80, use a heavier gauge metal, press many, many times, increasing the pressure just a little each time, or press once or twice with the 95 to establish a good flange, and then proceed slowly with the 80.

Always anneal metal before you begin forming. To achieve greater depth, you will need to anneal and press a number of times. It is important that the metal be put in the die accurately each time. Use tape to keep it in proper alignment. Keep track of how much pressure you are using and increase it in small increments after each annealing.

To remove the flange, shapes that have been formed in matrix dies must either be sawn out by hand or cut out with a matching blanking die. (See chapter V.) Shears will deform the edges.

embossing dies

22 Embossing is the process of forming metal in shallow relief. Embossed metal is formed on both sides whereas stamping or roll printing merely imprints a pattern on one side of the metal.

23 Flat and noncompressible found objects such as washers, lock nuts, metal mesh, small gears and keys may be used for embossing as can some textured plastics. Organic materials compress and rocks can shatter, thus neither should be used in the press.

Annealed metal must be used for embossing. Copper and silver from 28 to 36 gauge work best. The thinner the metal, the more detail that is possible. Metal heavier than 28 gauge does not emboss because its thickness prevents the metal from bending into small enough radii.

Use either ⅛-inch or 1/16-inch 95 pads (which are red) or ½-inch or 1/16-inch 80 durometer pads (which are yellow and yellow orange) with embossing dies.

22

23

24

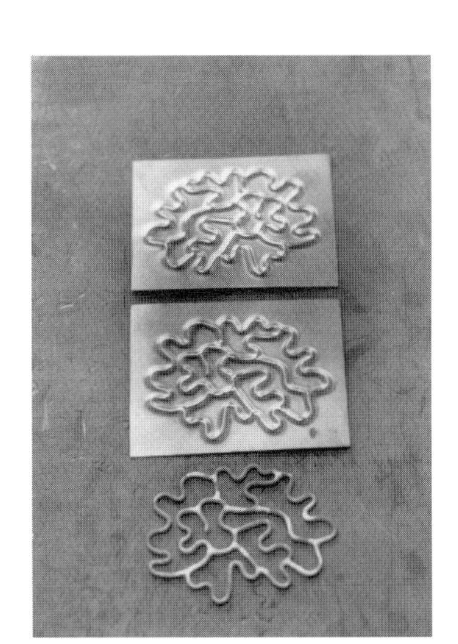

25

Small dies with a margin of ¼- to ½- inch around the impression area concentrate and maximize the hydraulic power that is available. The larger the die, the greater the area over which the pressure is distributed and the less effective the forming. Dies without a margin around the edges tend to have a deeper forming in the center of the die and fall off around the periphery.

Embossing dies are made in a variety of ways. **24** Wire patterns can be soldered to sheet to make a simple embossing die. **25** If a wire die is not soldered to a backing, it may be used from either side to make mirror images. **26** Matrix-type embossing dies can be made by piercing a pattern into a sheet of 16-gauge brass. **27** A shallow relief may be carved with burrs into acrylic sheet to make an *intaglio* embossing die. The effect is somewhat similar to repoussé, and suggests how embossing dies could be employed as a first step when using this technique. **28** Embossing dies can also be made by building up thin layers of acrylic or metal.

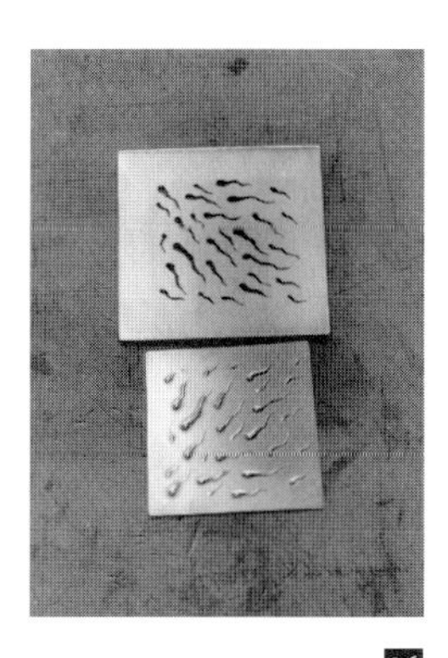

26

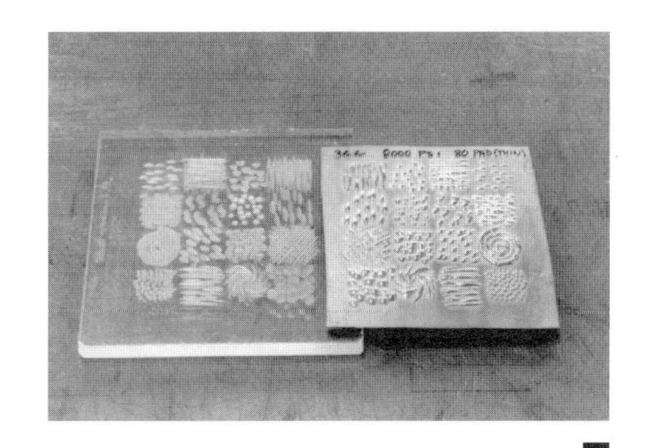

27

Always use one or more spacers when embossing so that you are working the ram in its mid-range where it is most efficient. Some experimentation is necessary as it is impossible to predict which pad or metal thickness will give the desired result. *Embossing often requires close to all the power that you have available.*

Pewter embosses with extraordinary results. The metal almost extrudes into the die, picking up an amazing amount of texture and detail. Unlike other metals, any thickness may be used and the back side of the pewter remains fairly flat. Considerably less pressure is required. When embossing pewter wait a while before releasing the pressure. It seems to have a memory!

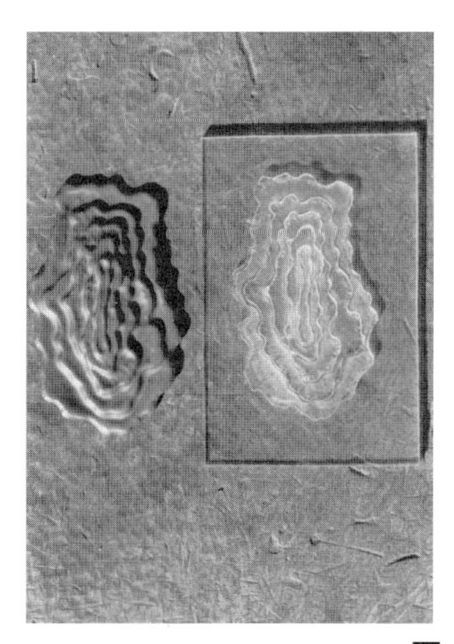

28

tips & troubleshooting

non-conforming dies

• What do you do if the metal tears? Use less pressure and stop to anneal sooner, use a heavier gauge metal and/or re-work the die in the places that are causing trouble. It usually requires some experimentation to determine how to use a die for the best possible result.

• What is wrong when only one side of the die "works" when the movable platen of the Bonny Doon visibly "tilts" during forming or the middle platen of the adjustable press gets "stuck" and doesn't go down? Problems such as these occur if anything is off-center. Your work must be placed on the center of the platen, directly over the jack and the ram of the jack must be centered directly under the bottom platen.

• What is wrong if the jack fails to supply power beyond a certain point? This may be due to only the top inch or so of the press being worked. You must use a spacer or two with thin matrix dies and with embossing dies. Spacers can go either under and/or over the die to create a metal and urethane "sandwich." Using spacers keeps the ram in the mid-range of its travel where it is most efficient.

• What is wrong if you use a matrix die and nothing happens? When using matrix dies remember to always put the metal you are forming between the urethane and die. It doesn't matter which way (facing up or down) the "sandwich" is put into the press.

• What is wrong if you use an embossing die and nothing happens? When using embossing dies, you must use one or more spacers, and make sure that the metal you are embossing is between the urethane and the die or found object.

• What causes uneven depth? Matrix dies that contain some wide and some narrow sections may form with uneven depth. Urethane will flow more easily and deeper into wider areas. A star with points, for instance, may form well and with some depth in the center, but the points may remain flat. In such cases emboss with a thin (1/16-inch or 1/8-inch) 80 or 95 pad first, anneal and then follow up with a 1/4-inch 95 (orange). You may then graduate to a thicker pad.

• How are well-defined edges maintained in a die? Over time, acrylic will wear and compress slightly. This usually isn't critical. To ensure well-defined edges on your matrix dies, you can face the die with 16- or 18-gauge brass or with 1/32-inch steel.

• What causes dies to crack? Cracked dies are caused by undercuts, by the cutout not being centered in the matrix die and by anything off center in the press. It can also be the result of accidentally leaving a tool bolted to the upper platen.

chapter III
conforming dies

Conforming dies are useful for forming deep or complex shapes that cannot be made with simple one-part dies. They consist of two corresponding rigid "male" and "female" parts between which metal is formed. It should be emphasized that the conforming dies in this book are for *forming*. They stretch and compress metal into three-dimensional forms, but do not in themselves reproduce texture and surface detail. Conforming dies, as described here for use in the hydraulic press, should not be considered a replacement for hardened steel dies used for coining and stamping in bigger, more powerful presses.

There are two basic types of conforming dies, standard dies and flangeless dies. 1 A standard die forms a shape which is surrounded by a flat flange which must either be cut away or incorporated into the final design. 2 A flangeless die forms metal over its entire face without a flange and permits the use of pre-cut blanks. Either type of conforming die can be made inexpensively with pourable steel-filled epoxy.

I have used two different epoxies to make dies — Devcon® Plastic Steel® Liquid (B) and Devcon® Aluminum Very Liquid (F-3). Although the mixing ratios of resin and hardener are different, both are two-component systems and work in the same way. Because the aluminum material is less viscous, it pours and flows into the die with fewer bubbles and theoretically allows greater detail. The Plastic Steel®, however, has a greater compressive strength and is considerably less expensive. In my opinion, Plastic Steel® is the best choice. There are several ways to minimize the bubble problem which will be discussed below.

1

2

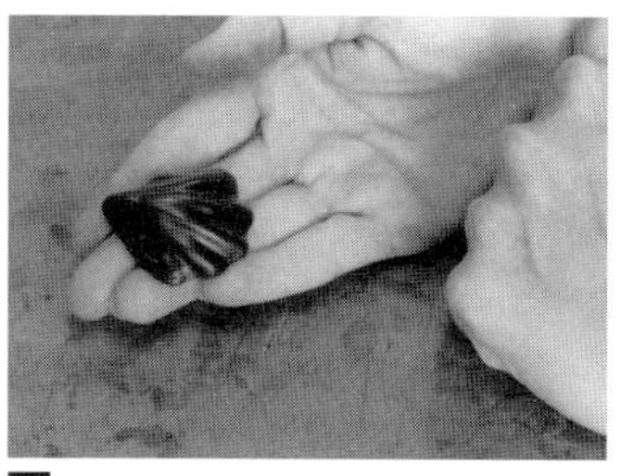

3

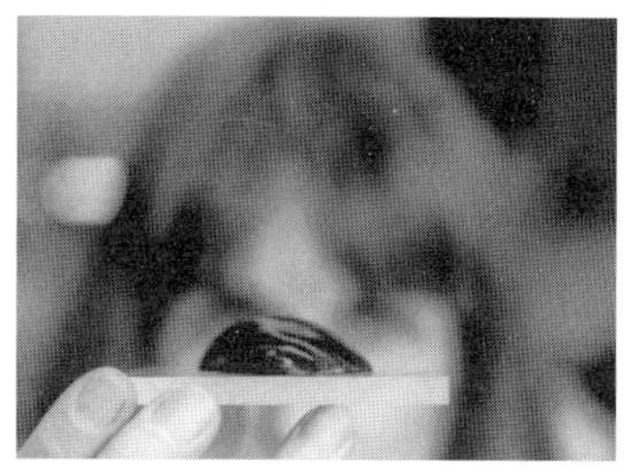

4

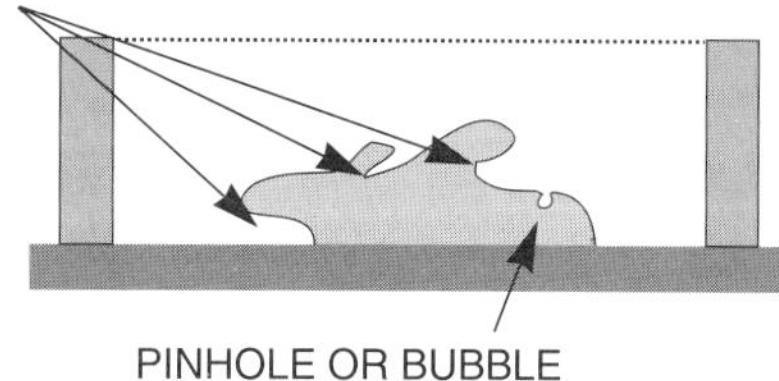

5

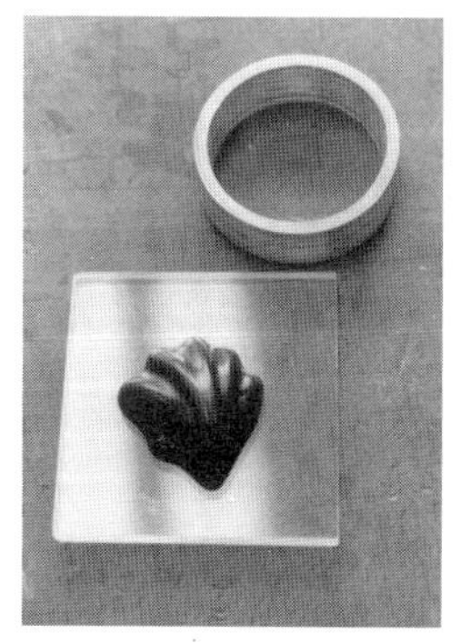

6

making the model

3 The first step in making a die is to create a model of the form you want to press. Acrylic, wax and plasticene models are better in many ways, but wood, plaster, metal or found objects can also be used. Exceptionally porous models must be sealed. Gentle curves and soft forms are more appropriate for this process than crisp, hard-edged geometric shapes. *Remember that conforming dies are for forming and that the fine detail, edges and textures of the model do not reproduce.* The resulting formed metal will reveal and obscure the model in much the same way that a thick coat of paint both reveals and obscures what is underneath.

4 When looking at the model from the side, there should be no undercuts, sharp angles or steep slopes and the highest area should be near the center of the die rather than near its outer edge. The height of the model is the same as the depth of the die and should be in proportion to the diameter. For example, a height (or depth) of ½ inch requires a die around 2 inches in diameter. Deeper dies should be proportionally wider. A 2 ½-inch diameter die could be ¾-inch deep. The nature of the relief must also be considered as each one is different. There is no exact rule.

5 Check the model very carefully for undercuts. Liquid epoxy will be poured over it and even a very small "lip" or "pinhole" will prevent the model from being removed. Soft materials such as plasticene and wax are more forgiving because you can "dig" the model out of the die and make modifications before pouring the second part.

6 Secure the model to a flat piece of acrylic that is at least one inch larger than the retaining ring. Plasticene and wax are self sticking and may also be used for bonding models made of other materials. I've yet to find an adhesive that I can recommend. Rubber cement can cause the edges of the model to be

gummy and irregular, some glues can be dissolved by solvents in the epoxy, super glues may be more permanent than is desirable, and double-sided tape leaves a small space for the epoxy to run under the edge.

Place a steel retaining ring around the model and plasticene around the outside to stick and seal it to the flat base. The ring serves to contain the steel epoxy when you pour it over the model and to prevent the die from cracking under pressure during use. When making a small, standard conforming die, the model can come as close as an ⅛ inch to the ring and within an ⅛ inch of its height. Larger and deeper dies require more space. But more than ¼ inch in either direction is unnecessary and wastes material. **7** When making a flangeless die, the model can fill the ring and be within an ⅛ inch of its height.

8 Retaining rings are simply short sections of steel pipe (⅛-inch wall minimum). They must be cut accurately at 90 degrees with flat and parallel surfaces. Round dies are preferred because of the convenience of using sections of round pipe as retaining rings and because a round die uses less epoxy than a square one. If the model is long and narrow, the rings can be modified by "squashing" them in the press. Tape the two rings that you intend to use for the top and bottom dies together, place them in the center of the press between the two steel platens and press until they are the desired shape.

Porous models (such as wood and plaster) must be sealed. **9** *Then brush the model and base inside the ring coated with release agent.* (If you are making a flangeless die, the base will not be exposed.) Devcon® Liquid Silicone Release Agent can be used, but do not allow it to "puddle." I have found that a thin coat of Vaseline® works very well. Plasticene models do not need to be coated because the epoxy will not stick to them.

7

8

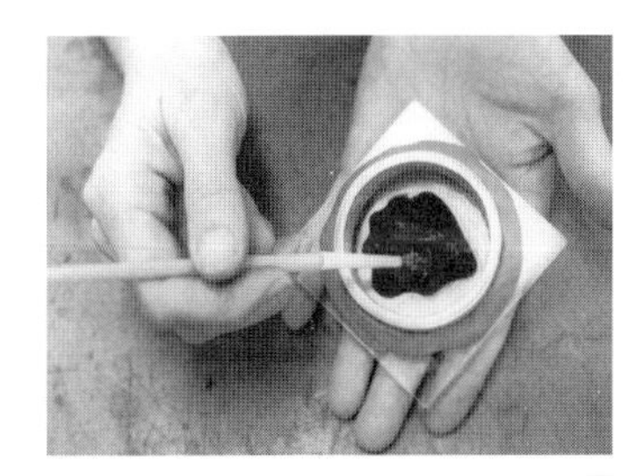

9

making the dies

Epoxy resins and hardeners contain materials which are irritants and are toxic. Almost 50 percent of industrial workers regularly exposed to epoxy develop allergies to them. I strongly urge you to read the Material Data Safety Sheets available from your supplier. If you decide to use Plastic Steel®, it is very important to take proper precautions.

Work in a room that is at least 60 degrees. Lower temperatures can adversely affect the viscosity of the material (making it too thick) as well as cause it to take longer to cure. Avoid skin contact with the solvents and hardeners and inhaling fumes. Wear goggles and gloves, and work with local exhaust or in front of a window exhaust fan.

10 Minimize your exposure time to these materials by being organized. Work time is approximately 45 minutes, but it is not unusual for the material to set up faster. Have your model complete, the release agent in place and quantities figured before you open the containers. Do not leave the containers standing open as you are working. Clean up immediately. Dispose of paper towels, mixing cups and gloves in a covered container or outside the room.

Plastic Steel® is supplied in quantities of one, four and 25 pounds. (One pound will make two or three sets of 2-inch conforming dies.) Our application requires mixing the components in small quantities. This must be done accurately to ensure good results and avoid waste. I recommend a gram scale and measuring by weight

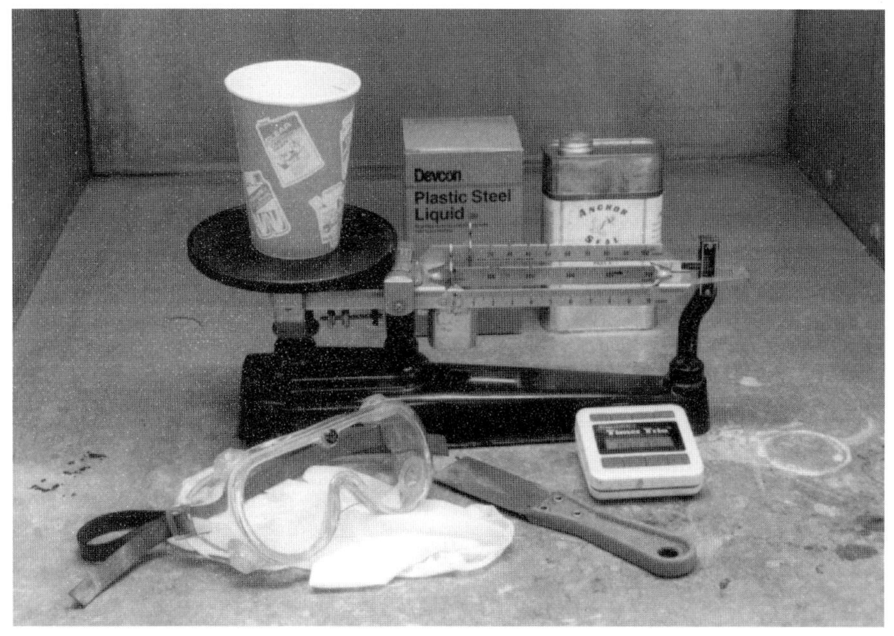

10

rather than by volume. Formulas for this are listed in the appendix (p. 80). There is also a table that lists the amounts needed for commonly used sizes of dies (p. 81). The numbers are rounded upward as there is always some loss in mixing.

Use a paper resin mixing cup, available at plastics stores, or a regular paper cup. Do not use styrofoam or plastic. Balance the gram scale for the weight of the cup, then set the scale for the weight of the resin. The resin, the thick black material in the can, must be stirred separately before weighing because the steel particles tend to settle on the bottom. You will need a metal spatula for this because it is so stiff. A 1-inch putty knife works well. (Screwdrivers are not satisfactory.) Add resin to the cup until the scale is balanced.

Next re-set the scale for the total weight of the epoxy and pour the clear liquid hardener slowly and carefully into the resin. Do not contaminate either container with the contents of the other. Mix thoroughly with the same metal spatula for at least four minutes by the clock, making sure to scrape the sides and bottom of the container. Try to avoid stirring air into the mixture as this causes bubbles in your die.

When you are ready to pour, the epoxy should be the consistency of pancake batter (without lumps). Although it is not usually necessary, Plastic Steel® can be thinned with a small amount of Anchor Seal Epoxy Thinner. Do not exceed 1 teaspoon thinner to 75 grams of epoxy, or 10 percent, and mix it thoroughly.

To prevent the formation of bubbles and to fill in the fine detail, you can "paint" it over the model with a small, soft, disposable brush. You must be careful not to brush away the release agent in the process. Then pour it in a slow stream into the lowest area of the die and let the Plastic Steel® "creep" over the model, filling to the top of the ring. The Plastic Steel® must completely fill the retaining ring. **11** I usually over fill slightly because even slight

11

12

13

underfilling causes problems. It can make the die difficult to use, and weaken it enough that it will crack.

Tapping the die helps "de-bubbleize" it. Placing the die on a vibrator of the type used to remove air bubbles from casting investment or placing it on a machine that vibrates when the motor is running will also help. Stop when the bubbles are no longer coming to the surface. Bubbles on the surface of the die are inconsequential and can be ignored. Finally, put the die in a safe *level* place away from your work area while it cures.

Hardening time for a ½-inch-thick die is about four hours, but a full cure takes 16 hours. Larger dies with a greater mass cure faster. Epoxy will not cure properly at temperatures below 60 degrees Fahrenheit. Place the die under a heat lamp or a regular light bulb to speed curing. (Keep in mind that if your model is wax or plasticene, too much heat will melt the model!)

When the epoxy is hard, remove the model from the matrix and make corrections in the die.

12 If the ring was overfilled, it should be filed so that the bottom of the die is even and flat. Conventional metalworking tools may be used to file and grind Plastic Steel®, but you must wear a dust mask and eye protection, and work only at low speeds. Clean up the dust immediately.

The matrix is used as a mold for pouring the punch so check again to ensure that you have no undercuts.

13 Small imperfections or even a bubble could prevent the dies from separating. Corrections can be made by either filing them away or by filling them with wax or epoxy steel filler. If larger repairs or revisions are necessary mix more Plastic Steel®. (The epoxy steel filler material available in tubes at most hardware stores is not as strong and cannot be reworked in the same way.) Epoxy will adhere to itself, providing the release agent is removed with Devcon® Cleaner/ Conditioner or another solvent.

Once the female part of the die is complete, you are ready to make the punch: the male part of the die. It is contained in another accurately cut steel ring. The punch should be the same diameter as the matrix but does not need to be as deep. A ½ inch is sufficient in most cases. *Put release agent on the surface of the steel rings where they go together and over the entire interior of the matrix die.* Do not coat inside the empty ring. **14** Place the ring over the matrix die and tape them together with masking tape.

Determine the volume and weight of epoxy needed to fill the die. Use the formula or consult the chart in the appendix. *This time additional Plastic Steel® is required to fill the space left by the model in the matrix as well as to fill the empty ring.* Estimate and add this extra amount. Follow the same directions for mixing, pouring and de-bubbleizing as before. Again, a heat lamp will speed and ensure a full cure.

Although the punch can be removed from the matrix after four hours, it is safer to wait 16 hours for the full cure before opening the die. Remove the masking tape and gently tap the die with a hammer. If it does not open easily, use a screw-driver to gently wedge the rings apart. If the die was overfilled, it must be filed or sanded flat. Small bubble holes in the punch can be ignored. Large imperfections can be repaired with more Plastic Steel® if necessary.

using conforming dies

Conforming dies work best with annealed malleable metals. Copper, silver, copper/silver alloys and the softer gold alloys are ideal. Pewter forms especially well in conforming dies and, unlike other metals, picks up a remarkable amount of surface detail. Niobium, titanium and (some alloys of) aluminum can be formed, depending upon the hardness as supplied. Brass forms, although not as easily. The best gauge to use

14

> Working with Plastic Steel®
>
> For small dies:
>
> Mixing time..............4 minutes
> Working time...........30-45 minutes
> Hardening time ....4 hours
> Curing time* ....16 hours
>
> (*room must be at least 60°F)
>
> REMEMBER TO USE RELEASE AGENT!

15

16

17

depends upon the size, depth and configuration of the individual die, and must be determined by experimentation. In general, larger, deeper and more radical forming requires metal 22 gauge or thicker. If the die is shallow and more detail is desired, then thinner metal should be used.

15 A wonderfully simple, low-tech method is used to align and secure conforming dies in the press. Apply double-sided Scotch® tape on the top and bottom of the die, put the die together and place it in the center of the press with the punch on top. Raise the platen until the punches are stuck to the platens. Then release the jack and allow the lower platen to drop several inches. That's it!

It's a good idea to test a conforming die the first time it is used and make necessary modifications. I usually do an initial pressing using 28 or 30 gauge copper. Cut a piece that allows a generous flange, at least ½ inch all the way around. Anneal, pickle and rinse the copper, but do not wipe it dry. Air dry, allowing the pink surface that can be rubbed off to remain on the copper.

Place the test piece on the female part of the die and raise the lower platen. Watch as the metal is drawn into the die. If the flange starts to fold, stop, take it out and flatten it with a leather mallet. This may be necessary several times. Stop when the dies are almost together. Excessive pressure can only damage the die.

Release the jack, allow the dies to part and remove the metal and the die. There may be folds and tears in the metal, but the quality of this first pressing is unimportant. Examine the die. 16 The pattern of "pink" that has rubbed off the copper will tell you where the die is tight. You should "relieve" the die in these areas by careful filing to allow more space for the metal.

If you plan to use the die to form metal thicker than 24 gauge, you may need to provide additional space in the die for the thickness of the metal. In a standard conforming die, you do this by filing the edge around the perimeter of the matrix and filing down the highest areas of the punch. Remove about the thickness of the metal you intend to use. The areas marked in the test described above are particularly important. In a flangeless die, file down the punch. Repeat the test a second time, if you wish.

Next, cut a metal blank the thickness you intend to use. Metal of 24 to 28 gauge can be used if the die has a shallow relief and greater detail is desired, but 18 to 22 gauge is needed if the die is deep, has steep angles, forms complex curves or will be used to make hollow forms. Preshaped blanks are used with flangeless dies. For standard dies, the minimum width for a flange is ⅜ inch. *Unlike matrix dies, conforming dies draw metal into the die,* so deeper dies and those with steep angles will likely require even wider flanges. Remember that the blank will need to be even larger than the die if the form is close to the retaining ring.

Again, using the double-sided tape, align the dies in the press with the matrix on the bottom.

17 Place the "blank" between the dies and raise the platen, watching what happens to the flange as the dies come together. 18 If folds begin to develop, take the metal out and flatten the flange with a leather mallet, anneal and resume pressing. Depending upon the die, it may be necessary to repeat this process several times before the dies come completely together. 19 It is not necessary and not advisable to apply additional pressure once the dies are together because you can seriously damage or even break the die. Write on your samples how you did them, the gauge of metal and the number of annealings.

If the die is steep or deep and you have problems with excessive folding of the flanges, pre-forming with urethane will eliminate this problem. 20 To pre-form, put an annealed blank in the smallest contained block in which the die will fit, place the female part of the die over it, and press. If a pusher is necessary, it must be at least as wide as the die.

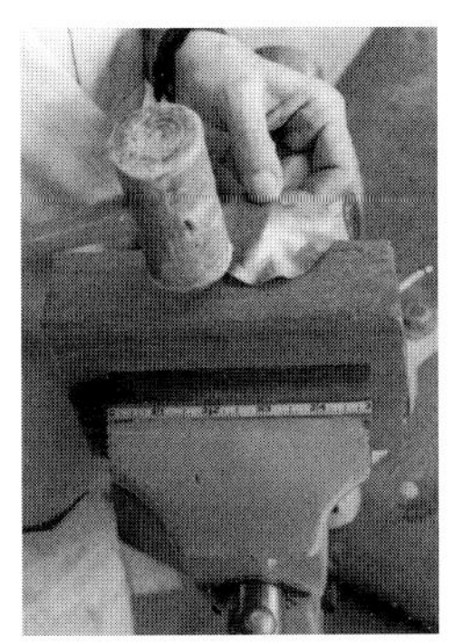

18

19

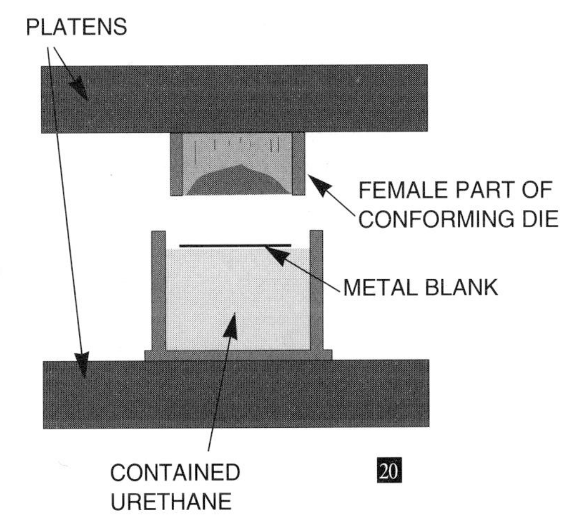

20

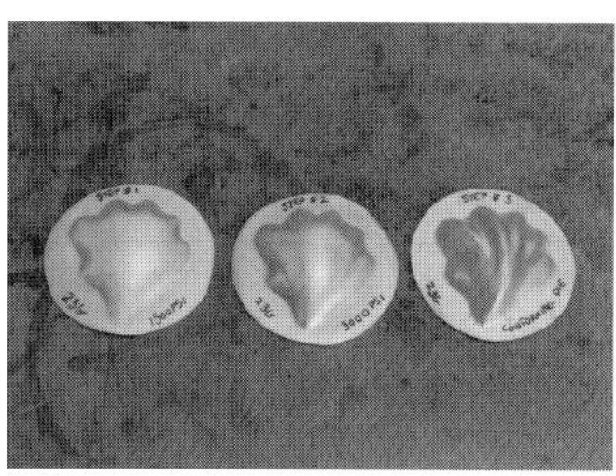

21

22

23

21 Pre-forming establishes a flat flange and stretches metal evenly into the die. Deep dies may require two or more pressings into urethane before the two rigid parts of the conforming die may be used together. Again, keep a record by making notes on the samples.

Each die is different and experimentation is necessary. If the metal tears, anneal and press in smaller stages or use a heavier gauge. The number of annealings and whether the urethane pre-forming step is necessary will depend upon the particular die. If there is an area that continues to cause problems it may be necessary to modify the die by filing it down.

22 The edges of the female part of the die will wear with use. If you plan to make a great number of pressings or if you need accurate sharp edges for making hollow forms, you can add a face plate. Cut the silhouette out of 18-gauge brass or steel and place this over the female part of the die. It can be held in place by glue or tape. A face plate will not affect the form when the die is used.

23 Finally, it is possible to use urethane to intensify conforming dies in order to attain maximum detail. After the forming is complete, anneal and place the blank back into the female part of the die. Then place a 1/16-inch 80 durometer urethane pad between the punch and metal and apply moderate pressure. As with embossing, this works best when the metal is thin.

Of the dies described in this book, conforming dies are the most labor intensive and expensive to make. They also may require more testing in the press and more "fiddling" to get a satisfactory result. It is advisable to look carefully at your design concept to see if it can be accomplished in a more direct way. When it can't, conforming dies are rewarding and worth the effort.

chapter IV
blanking dies

Steel blanking dies used in manufacturing consist of a positive (punch) in the shape to be cut out and a negative die (matrix) with its silhouette. They are usually made with some mechanism for removing or "stripping" the blanks from the die. Perfect alignment is maintained by design in the industrial presses in which they are used. Certain types of blanking dies may be used in the hydraulic press, but the dies must be professionally made and are expensive.

1 An alternative for blanking that is consistent with the low-tech, low-cost options of the hydraulic press is the "pancake" die, made with a single angled cut in a sheet of tool steel. **2** The die forms its own hinge, ensuring perfect alignment. The concept was used in the 1930s for blanking sheet metal parts for airplanes and is similar to the "Continental Process" in its employment of an angled cut, producing the punch and die simultaneously. Roger Taylor developed a method for blanking using the same principle which he markets in England as the "RT Blanking System." More recently, David Shelton published information on the process based on his own research (*Metalsmith,* Volume

1

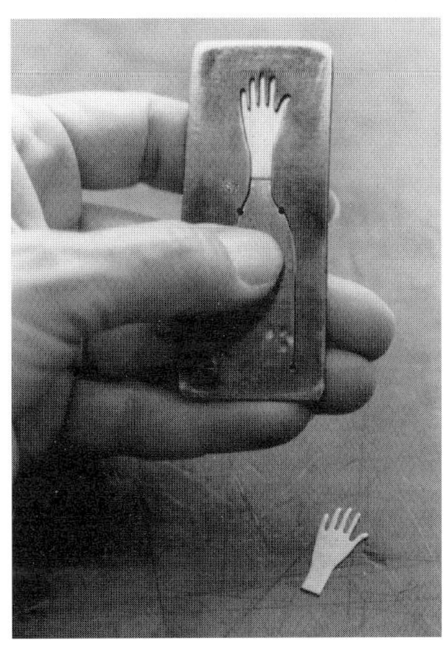

Die by David Shelton **2**

13, Number 1, Winter 1993). His expertise has grown into a die-making and heat-treating business which is listed in the appendix along with suppliers.

The effectiveness of the die depends upon three factors: (1) the thickness of the die, (2) the width of the saw blade, and (3) the angle of the sawcut. **3** Cutaway views illustrate the principle of this type of die and show how it works. The punch (positive) is pushed down, metal is inserted and even pressure is applied. The cutting edges of the die are kept in alignment by the hinge which connects them and the metal is sheared as the cutting edges pass each other.

Pancake dies have been used in a 20-ton press to cut nonferrous metal as thick as 14 gauge. A limiting factor, other than the thickness of the metal, is the complexity or length of the cutting edges. Lengthy cuts may require more force. Pancake dies will shear not only flat sheet, but can be used to blank metal that has been formed in matrix dies. This is explained in chapter V.

While mild steel can be used for short runs, high carbon steel is a better choice. If made and used correctly, a tool steel pancake die will produce hundreds of blanks. Tool steel called "oil hardening precision ground flat stock" is recommended even if you do not heat treat it. Suppliers are listed in the appendix.

If a longer wearing die is needed, tool steel dies can be heat treated. When this is done correctly, the die could last as much as 20 times longer. The process involves heating the die to a specific temperature, quenching it in oil and then tempering it to the appropriate hardness. This process, developed by David Shelton, will be described.

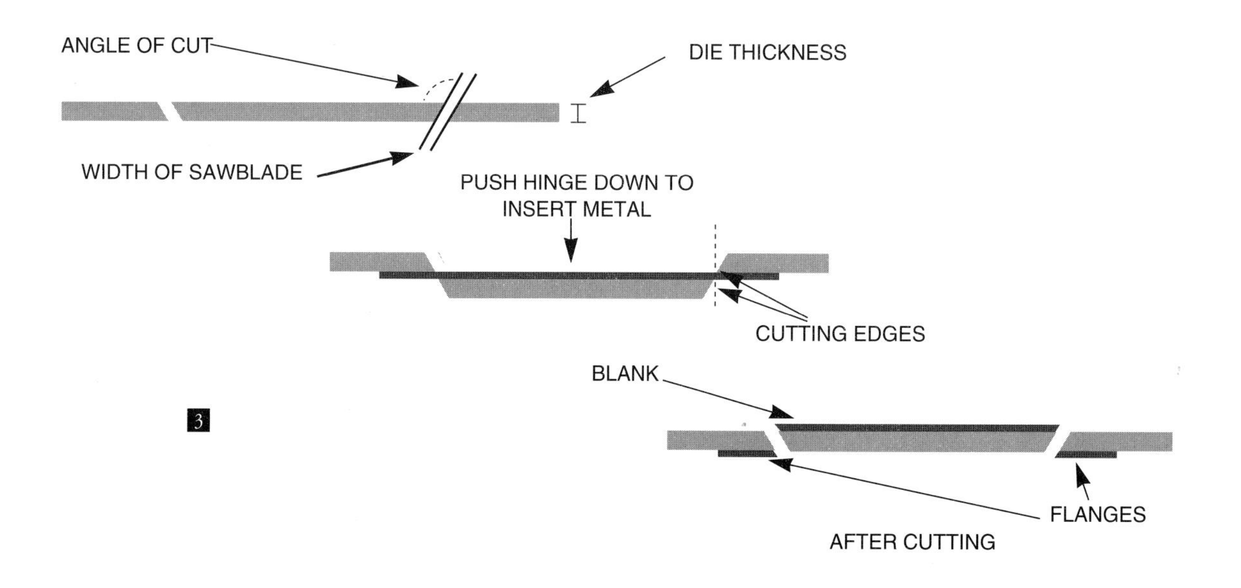

layout and sawing

There are a number of variations to consider when laying out a die. Note that the hinge leaves a flat edge or tab. [4] Normally this is placed where it can be trimmed most easily, but occasionally the tab can be made part of the design. [5] Complex shapes can be made with two or more hinges. Multiple hinges can be used to cut several blanks at once. [6] Another alternative is to make a die to trim edges rather than to blank out shapes. [7] Small designs can be ganged in one die, connected by tabs, or [8] dies can be laid out to cut multiple shapes from strips of metal with a minimum of waste. In either case, there must be at least ⅜ inch between the shapes. [9] With advance planning there is little waste.

Pancake dies require careful planning and precise marking. Because each die is different, it is impossible to specify measurements for the length of the hinge or the distance between the holes at its base. The hinge needs to be long enough to spring open for

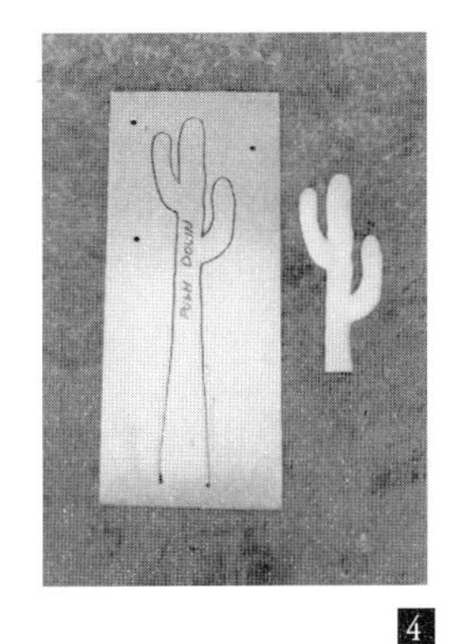

4

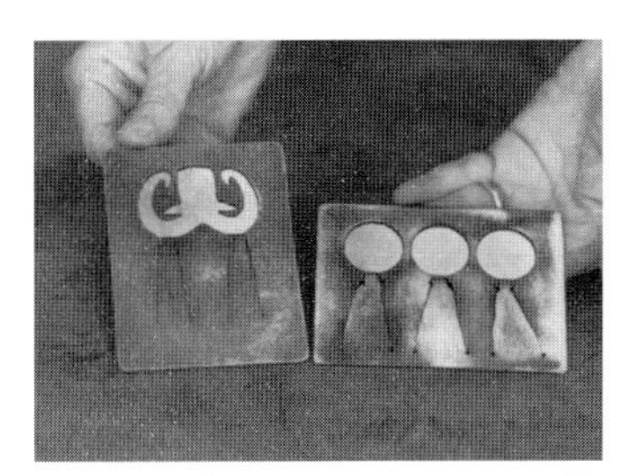

Dies by David Shelton 5

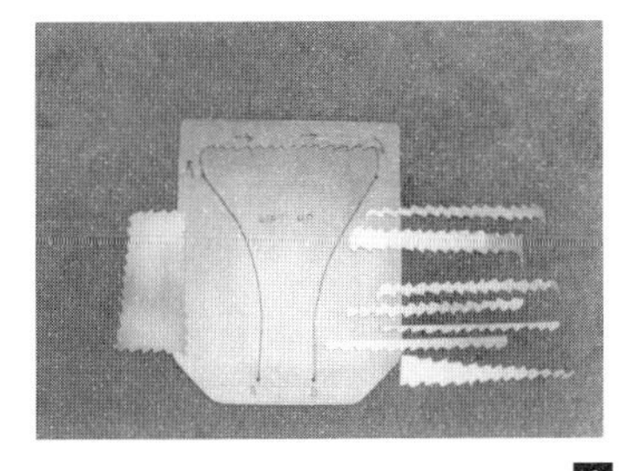

6

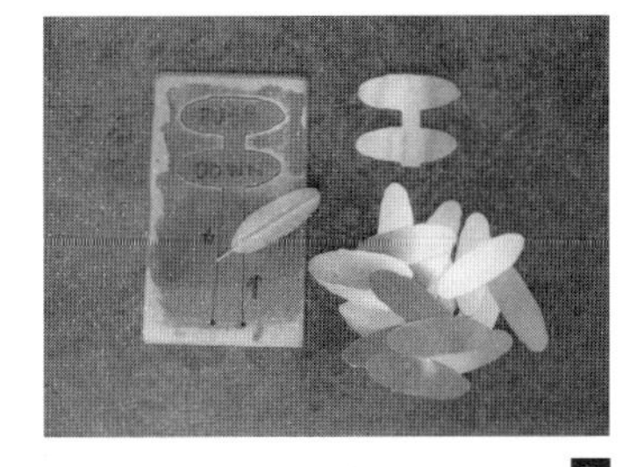

7

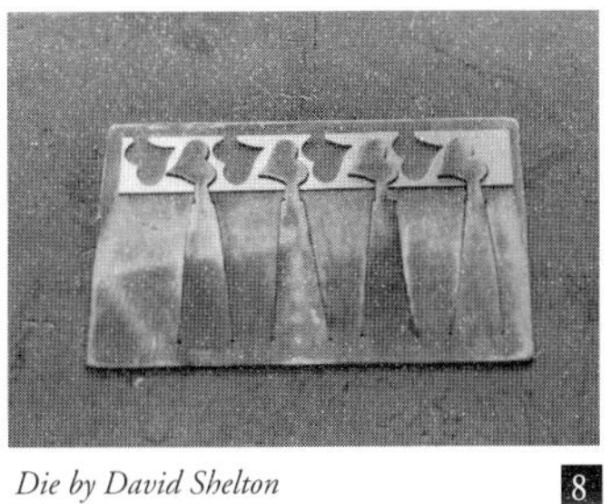

Die by David Shelton 8

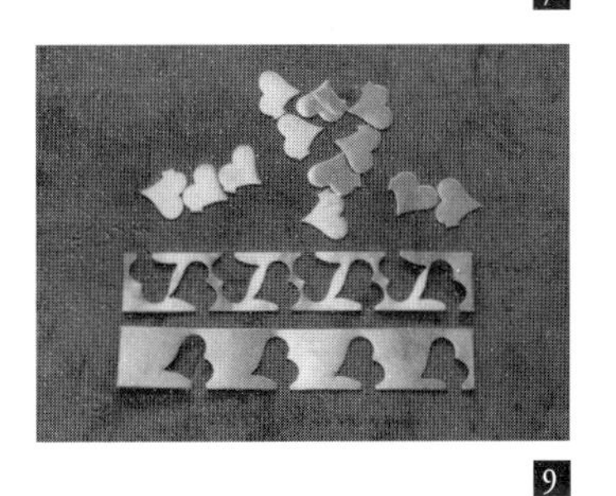

9

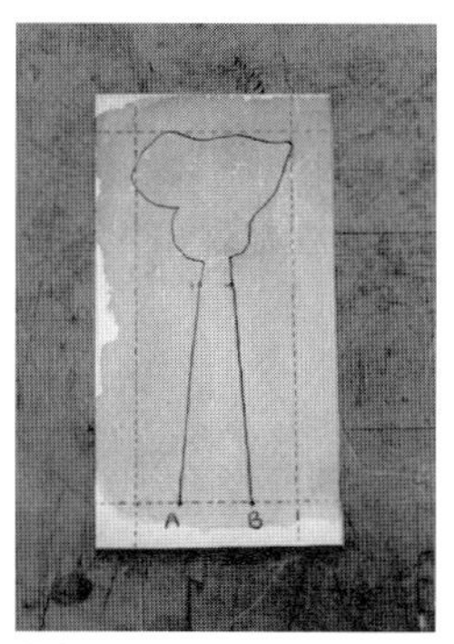

10

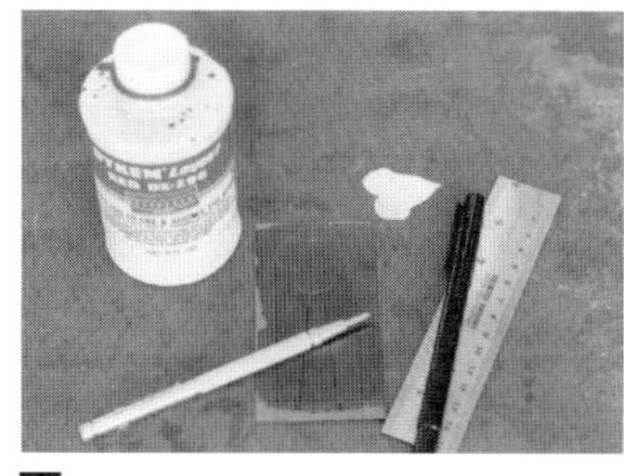

11

inserting metal, not so narrow that it twists, causing the die to shift out of alignment and yet not too wide (especially in thicker steel) to flex. 10 As a general rule, the length of the hinge should be at least 1 ½ times the width of the blank. For dies 1/32- and 3/64- inch thick, allow for a margin of 3/8-inch between the sawcut and the edge of the die. A ½-inch margin is necessary for thicker dies (1/16 and 5/64). 11 For accuracy, use layout liquid and scribe the design on to the steel plate .

Two sets of holes for each hinge are another option. (5, 8) The idea is to place one set of holes at the bottom of the hinge and an additional set of holes closer to the shape to be blanked. The two long saw cuts that make the hinge are not cutting edges, do not need to be angled and can be accomplished with a larger sawblade. This will save some time and energy. However, remember that when putting metal into the die, the cutting edge of the die stops at these holes which may make it inconvenient to use. After drilling the holes, be sure to remove any burrs that exist on the back of the die. If the die does not lie flat on the sawing table or bench pin as you are sawing, the angle can be thrown off enough to cause problems.

The thickness of the metal to be blanked determines the thickness of the tool steel to be used. The die must be at least as thick as the metal to be blanked, and it is better to make it slightly thicker. Intricate designs should be made of thicker steel to prevent small projections from becoming bent and misaligned. A table in the appendix (p. 82) lists recommended die thicknesses for various gauge metals and the corresponding sawblade sizes and angles.

The sawblade size is chosen in relationship to the particulars of the die. Intricate patterns require the use of smaller saw blades to make tighter turns. Larger blades cut faster and make the job easier but necessitate sharper angles. Sharp angles result in cutting edges that wear down quickly unless the die is heat treated.

The angle must be determined carefully. The angles listed in the table were determined mathematically and adjusted. Sawblades, due to the set of their teeth, cut a slightly wider kerf than their measured width. Because there is also some variation among different brands of sawblades, it may take some practice before you find the ideal combination of blade and angle.

A "loose" die (with insufficient angle) will produce a burr and may cause metal to stick in the die, especially when blanking thin metal. A tight die made with too sharp of an angle may not open and may wear out prematurely.

Also consider whether or not the die will be heat treated and the method of sawing. If the die will not be heat treated, use the angles in the table. In this case, it is better if the angles and edges are not too sharp and the die is slightly loose.

If a die is to be heat treated and sawing is done with a sawing guide or power saw, increase the angle by one degree over what is listed in the table. Heat treated cutting edges are considerably harder and will last a long time. *The tighter the die is made, the cleaner it will cut.*

On the other hand, if the die is sawn free hand with an angled bench pin, increasing the angle is not advisable. The inevitable variations in the angle can prevent the die from opening and/or cause it to stick.

While it is not absolutely necessary to have special equipment for sawing, the Bonny Doon precision sawing guide is highly recommended for its exceptional design and ease of use. By keeping the blade perfectly vertical, it allows you to use a steeper angle and make tighter dies. A scroll saw is another alternative for achieving an accurate and consistent angle, but it requires patience and practice to determine the correct blade tension and pressure, and to gain the fine control that is necessary for the accurate sawing of dies.

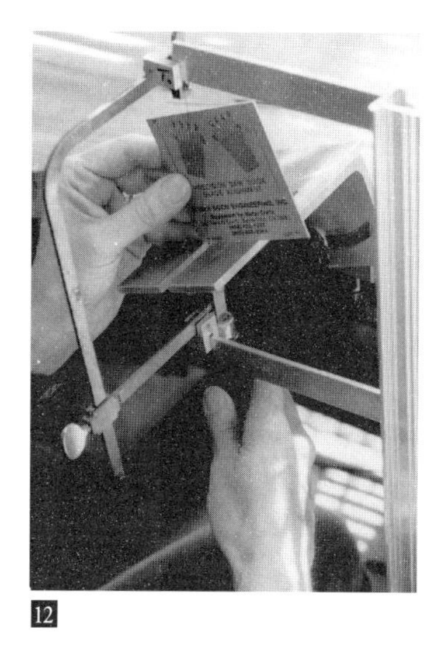

12

13

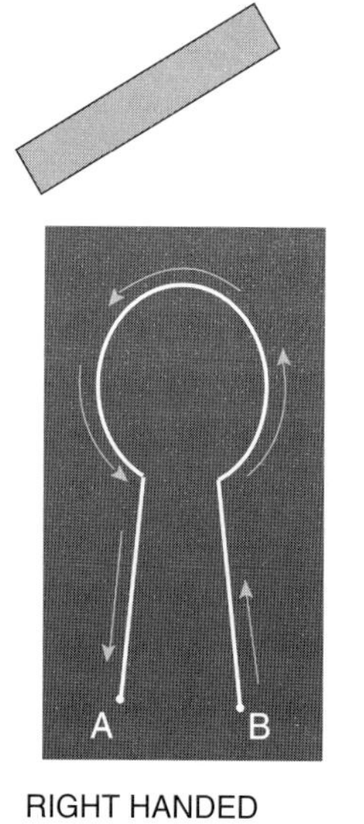

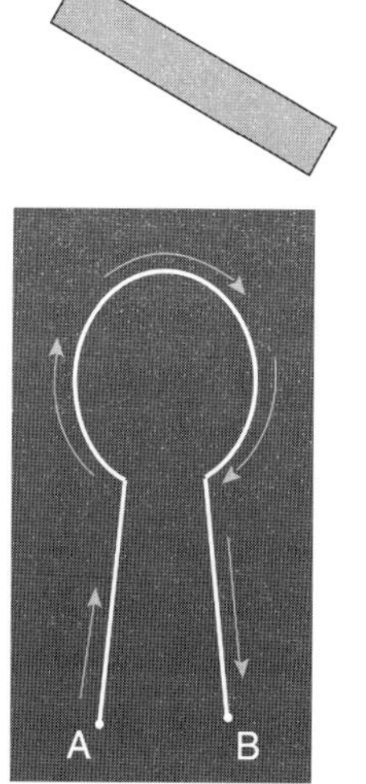

14 RIGHT HANDED DIRECTION LEFT HANDED DIRECTION

12 The sawing table or bench pin must be set at the angle that you have selected. 13 For right-handed people, the bench pin should slant down to the left. Begin sawing at point B and make one continuous cut, counterclockwise, ending at point A. 14 The slope of the bench pin and the sawing direction are shown in the diagram below. Left-handed people should do just the opposite. Angle the bench pin down to the right, begin sawing at point A and continue sawing clockwise, ending at point B. If you are using a scroll saw, the table should be angled to the left. Because the blade is facing you, you must saw from point A to point B to get the correct angle.

15 *A die with more than one hinge for a single shape requires that one direction be maintained in the sawing.* It will be clockwise or counterclockwise, depending upon whether you are right- or left-handed, but you must continue around the outline of the blank in the same direction. The arrows in the illustrations show the correct method (for a right-handed person).

If you are sawing free hand, it is important to saw straight up and down. You can't do this if you are sitting in an awkward position. Place your chair squarely in front of the bench with your shoulders level. Saw with motion from your elbow. Lubricate the sawblade often and use light even pressure. It takes practice and concentration.

The precision sawing guide takes the "wobble" out, but you must still determine the optimum blade tension and pressure. As you saw, keep both the metal and the blade moving.

When the sawblade breaks, you must thread it through the hole where you started and move it around the die in the same direction. When you have finished sawing, grind or sand to remove burrs and round the edges of the die. This will prevent the metal left after blanking from being imprinted with the edge of the die. Then mark the right side of the die, the side that was "up" as you were sawing, and write "push down" inside the cut.

If you are heat treating a die, do not open it until after it has been tempered. Opening and breaking in heat-treated dies is described on page 63.

15

If you are not heat treating, the die should open easily and be ready for use. If it will not open when you push down on the cutout, try to determine where it is sticking. Hold the die up to the light and mark along the edge of the punch where you cannot see through the kerf. If the die was sawn free hand it is probably because the angle is slightly off. This can usually be corrected by carefully filing the cutting edges until they clear, as shown in the illustration on page 54.

16 To keep the die open so that you can work on it, open it backwards by pushing up on the cutout and insert a strip of metal in the hinge.

16

using blanking dies

Before using a die in the press make sure the jack is centered and the plates come together evenly. To use the die, push the punch, the center part, down and insert the metal. Make sure that the metal does not extend below the cutting edges if you have two sets of holes in the hinge. To prevent the steel plates from scratching your metal, place the die and metal between two pieces of acrylic. *Do not use urethane pads with blanking dies!* Place the die in the press in such a way that it is centered and fully supported by the acrylic. Press until you hear a loud "pop." You should find a cleanly cut blank separated from the flange.

If you feel resistance but hear no sound, do not increase the pressure. Stop to see what happened. Sometimes, especially when the material is thin, there is no pop. *If the metal is completely stuck in the die or embosses rather than cuts, it was most likely put in backwards.* (You didn't "push down" from the front side to insert the metal.) Avoid this mistake. It may cause the die to become misaligned. (If this is happens, follow the directions for realigning heat-treated dies.) If the die sticks in just one area, it may be that the plates of the press are not level or that the sawing angle was not maintained. In the latter case, there is no way to correct it but you may still be able to use the die to blank thicker metal.

16

heat treating dies

The method described here for heat treating dies was developed over a period of years by David Shelton, who claims to have also fully explored many heat-treating avenues that were unsuccessful. 17 You will need a small kiln with an accurate pyrometer, safety glasses, leather gloves, common pliers and quenching oil. (Because motor oil contains additives and impurities which may be toxic, it is inadvisable to use it for quenching.)

The first step after sawing the die, de-burring and grinding the edges, is to harden it. Set the pyrometer to 1500° F (815° C) and allow the kiln to heat. (It is very important that the pyrometer is accurate.)

Once the kiln is up to temperature, place one or more dies in it. Do not put dies on the floor of the kiln. Each die must be fully supported by a kiln shelf. Dies may be stacked up to 3 high as long as they are fully supported. Place each die with the hinge end closest to the door in such a way that it can be removed quickly with pliers or a vice-grip.

It will take 10 minutes or longer for a die to be evenly heated to the correct temperature. When it has reached 1500° F the die should glow the same orange-red color as the walls of the kiln and have no dark, cooler areas.

The die must be removed and immediately quenched in warm (125° -175° F or 52° - 79° C) oil. **18** Grip the die firmly near the center of the bottom edge, taking care not to squeeze too tightly or far into the hinge. The die must be perfectly vertical as it is plunged into the oil, then moved slowly around to promote even cooling.

When it is cool enough to be held, it may be removed and cleaned. The die at this stage is extremely hard and very brittle. If the die is warped or if the hinge is out of alignment, resist the temptation to correct it at this point. The appropriate time occurs later.

The next step is to *temper* the die. In this process, the die is gently reheated in the kiln to a predetermined temperature which reduces its brittleness and makes it more flexible, yet retains enough hardness to stay sharp. Generally, 550° F (287° C) is recommended. The temperature range is 450° - 600°F (232° - 315°C). The lower temperatures leave the die harder and thus more suitable for use with metals such as 14K gold or nickel silver, but also more brittle and likely to break. Tempering in the upper range makes a stronger and more flexible

18

19

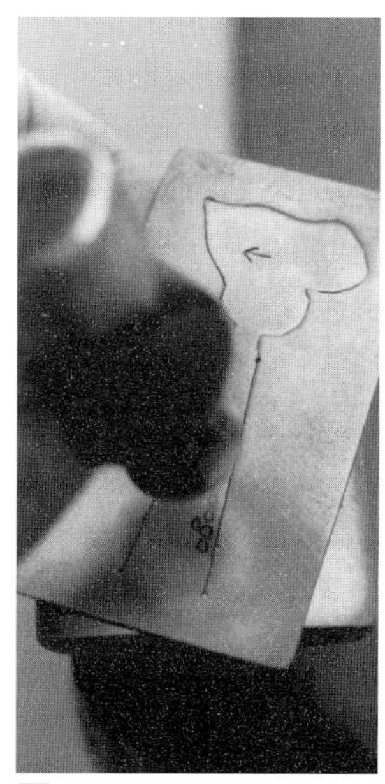

20

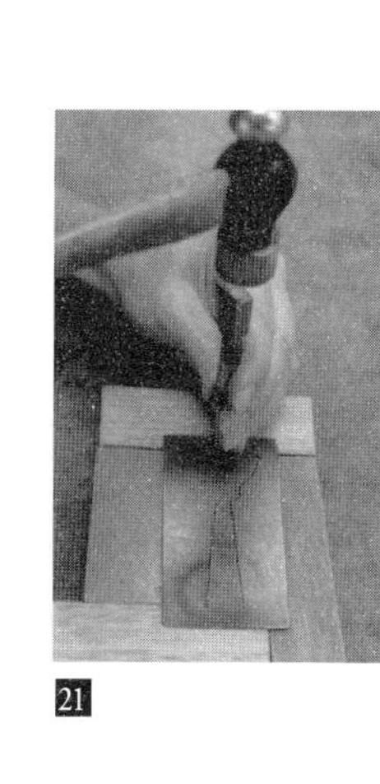

21

22 *Die by David Shelton*

die, but one that may wear out sooner. Size, detail and intended use should be considered when deciding on the temper. Experimentation may be necessary.

Load the die or dies back into the kiln and re-set the pyrometer. The dies should be "soaked" for one hour at the desired temperature and allowed to cool slowly. Ideally, turn the kiln off and leave it overnight.

After cleaning the die again, you can anneal the hinge so that it has more spring. 19 Use a torch with a small flame to heat the hinge and the area adjacent to its base until it is light blue in color. If there are any delicate parts that might be more vulnerable to breaking, they should also be annealed at this time. This would include tabs between multiple parts or narrow necks in the punch. Then let the die air cool.

Next, hold the die up to the light and inspect the alignment. The kerf should look even all the way around. This is the time to make adjustments if the positive element (punch) has shifted. 20 The hinge can be stretched and realigned with a few well-placed

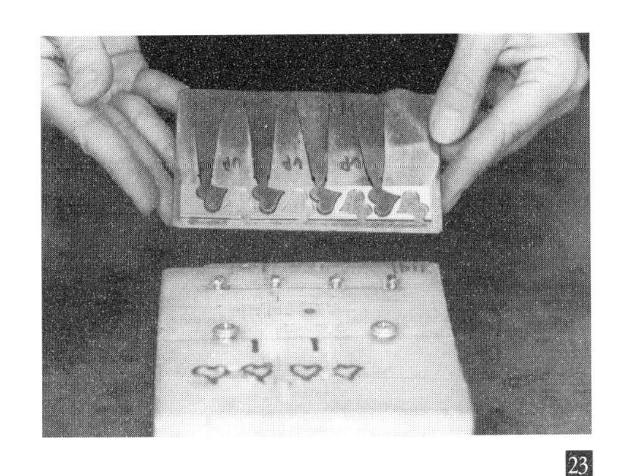

23

Die and loading block by David Shelton

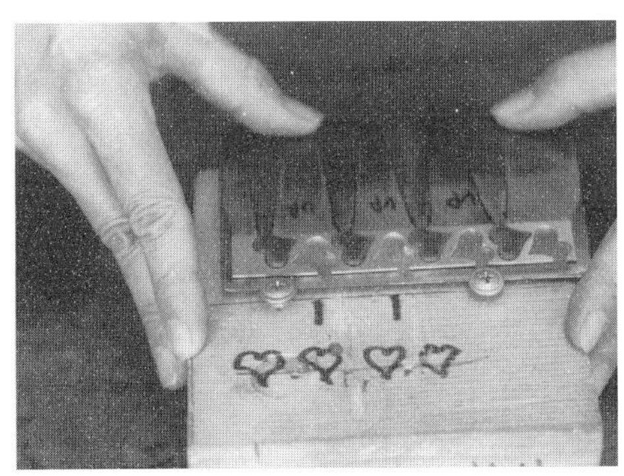

24

blows of a ball peen hammer. This is done on an anvil and from the back side of the die. The punch will move in the opposite direction of the blows as shown in the illustration. Misalignment causes the metal to jam in the die and can, if not corrected, ruin the die. Dies should be inspected during use and, when necessary, realigned in the same way.

Sometimes, especially when you have intentionally made a tight die for blanking thin material, it may not open easily. 21 If this occurs, break it in by forcing the punch down the first few times it is used. With the edges of the die supported, tap the positive down with a flat-ended punch. Start at the top as shown in the illustration.

Even then, large thick dies and those with multiple hinges can be difficult to open and inconvenient to use. 22 It is possible to pre-spring them slightly by re-annealing while in the open position. Slide metal into the die to open it and heat just the section at the base of the hinge between the two holes. 23 Another way to solve this problem is to make a loading block to open the hinge or hinges. 24 *The die is used upside down.* A pair of screws in the loading block hold the top edge of the die down and a screw lifts each punch as the bottom edge of the die is pushed down.

blanking dies

• **What is wrong when the metal sticks in the die?** If the die was sawn "free-hand," the angle is probably off and can be corrected by filing. Metal also sticks in the die when it is loose, misaligned or worn out. If it is simply loose, you may still be able to use the die with thicker metal. If it is misaligned, you may be able to straighten it by hammering the hinge. If it is worn out, you must start over.

• **What is wrong when the metal parts have a burr?** This is also caused when the die is loose, misaligned or worn out. (See above.)

• **What causes marks or scars on the blanked metal parts?** Sharp corners and edges on the die or rough unfinished steel platens on the press can cause this. The edges of the die should be rounded before heat treating. Acrylic spacers in the press can be used to protect the metal.

• **What causes dies to break during use?** This can be caused by the platens of the press not being parallel, by not placing the die in the middle of the press, by leaving scraps of metal in the press, by not supporting the die fully in the press or by using excessive pressure. Breakage can also be traced to the heat-treating process. The tempering may have been at too low a temperature or for too short a time.

• **Is something wrong if the die"clicks"?** This happens when the die is tight and is okay as long as you can open it with your fingers.

• **What is wrong if the punch shifts evenly out of alignment?** This is not unusual. Hammer to adjust.

• **What went wrong if the punch has shifted unevenly?** This could have happened during the hardening process as a result of overheating or of uneven heating or cooling. It could also be caused by uneven heating while annealing the hinge. You may be able to save it by selectively annealing areas that need to be stretched and giving them a few well-placed blows with a ball peen hammer.

• **What is wrong when the die warps during heat treating?** This can be caused by uneven heating in the kiln or by not holding the die vertical when quenching in oil.

• **What causes dies to crack during heat treatment?** Cracking can be caused by the quenching oil being too cold or water being mixed in the oil.

• **What causes dies to wear out prematurely?** This is due to the steel being too "soft." It could be a result of the die not being hot enough when it was quenched or being too hot when it was tempered.

chapter V
die variations and combinations

1

2

Dies can be used in different combinations and along with other metalsmithing techniques. Once you become familiar with the basic principles of making and using the various types of dies described in earlier chapters it is hoped that you will be able to combine their use with other metalworking processes. The following suggestions are just a few of the many ways in which one might do this.

Interesting results are achieved when metal that has been previously manipulated is formed into matrix dies. 1 For instance, folded and fold-formed metal can be used in matrix dies following the sequence: anneal, fold, anneal, unfold, anneal, then press with urethane into a matrix die. Folds can be left soft or made sharp by hammering or flattening them in a rolling mill. Different thicknesses of metal yield different results.

2 Embossed metal can be formed in a matrix die. It is usually better *not* to anneal the metal after embossing, and to place it so that the urethane presses the *back* of the embossed sheet into the die so as not

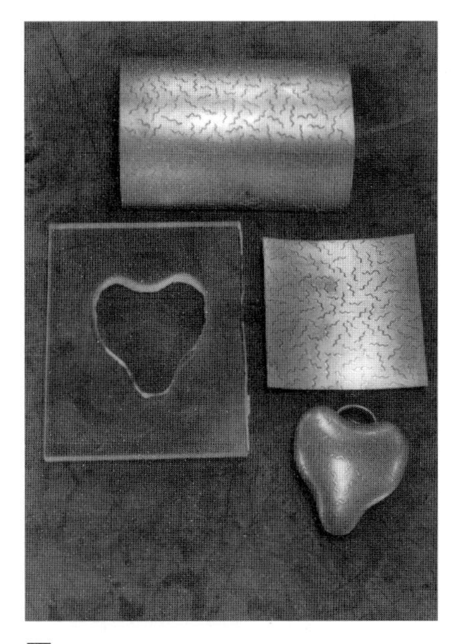

3

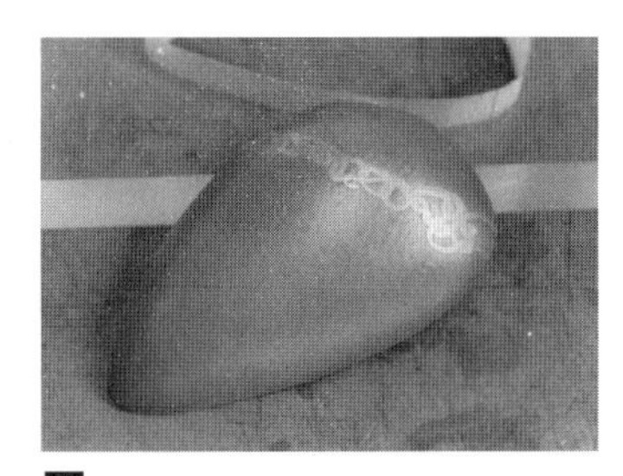

4

5

6

7

to flatten it. Flattening it, on the other hand, can create interesting effects.

Metal that has been previously "processed" can be formed in matrix dies. 3 Although there will be some distortion from changing a flat surface to one that is contoured, roll printed metal can be formed without damage to the surface. 4 Metal with soldered inlay or overlays can be formed into matrix dies, as can mokumé. 5 A flange and blank cut by a pancake die can be used to make married metals and shaped into a matrix die. Because the forming takes place over the whole surface of the material at one time, the layers or seams joining different metals are less likely to split apart. Depth should be achieved very gradually with multiple annealings. Etched metal can be formed into matrix dies as long as the depth of the etch is taken into consideration. If the etched areas are too thin, they will split open as the metal is stretched into the die.

Matrix dies lend themselves to variations. 6 A **multiple matrix die** is simply a grouping of several cutouts on one die, enabling you to press several shapes at the same time. 7 A **split matrix die** has two or more parts that can be aligned in various ways for different effects. Round matrix dies can be made for use with contained urethane for maximum efficiency. A round matrix die with a face plate is shown on page 33.

8 A matrix die can be combined with an embossing die. The result is a raised pillowed form with detail in the flange. The embossed part, may be added to the surface of the matrix, as shown in the illustration, or cut into it. For the best results, first emboss with a thin pad, then use a thicker pad for achieving depth in the matrix.

9 An "**insert**" can be used with a matrix die to add detail to the "pillowed" form. To add a linear detail, solder thick sheet metal on edge, perpendicular to a backing sheet. It should be curved or angled for strength, and the backing sheet should be at least as large as the matrix die in order to provide full support for it.

10 In the example, three pressings were necessary to obtain the full effect of a curved line and a deeply pillowed form.

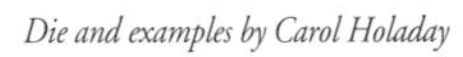

Die and examples by Carol Holaday

9

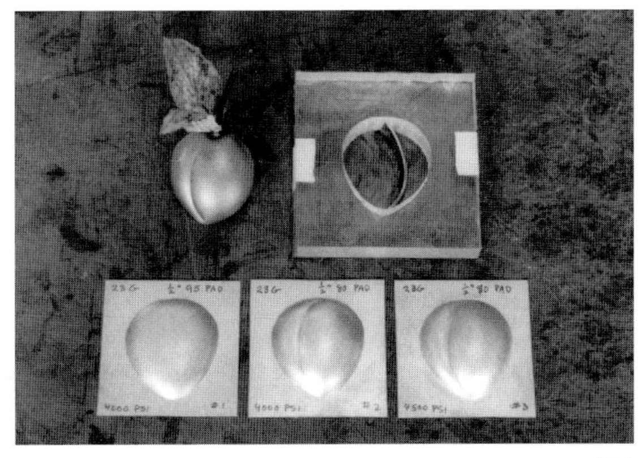

10

An embossing die can be backed up with a punch. 11 For instance, an embossing die made of wire soldered to a flat sheet is domed with a dapping punch or mushroom forming tool. The domed die is then taped to the punch, and the two are used together.

Embossing dies can be used in various ways in combination with mushroom formers, urethane and steel cups. (Remember to protect the mushroom with a brass face plate.) 12 The illustration shows how the simplest of dies, a coiled wire, can be layered with a metal blank and thin urethane pad to achieve different effects. In one, the wire spiral appears to be pressed into the dome, and in the other, the spiral is raised from the dome.

A blanking die and a steel stamping die can be used in sequence, offering more possibilities. Stamping linear detail into metal is quite possible in a 20-ton press as long as the die is small and harder than the metal being stamped. Low-cost steel stamping dies can be made by a process called Electrical Discharge Machining (EDM).

11

12

13

Blanking die by David Shelton
EDM die by Quicksilver

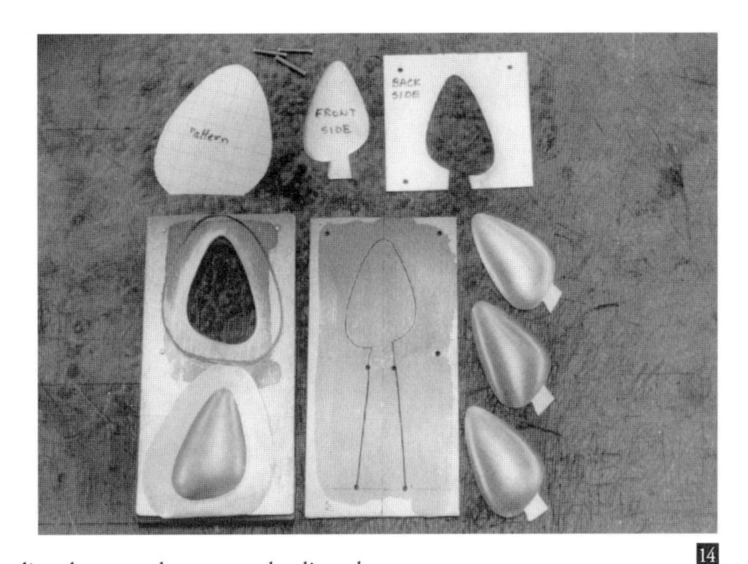

14

13 In the illustration, the flange left after blanking was filed slightly larger and taped to an EDM die for use as a placement guide. A company that makes custom stamping dies is listed in the appendix.

blanking/matrix dies

14 With advance planning and by following the instructions step-by step, it is possible to make a matrix die and blanking die that work in sequence, enabling you to cut out a shape *after* it has been formed. The blanking die is made first, and the matrix die is made to fit it by using the flange as a template.

The matrix die, in addition to being used to form the metal, is then used as a spacer during blanking, preventing the formed metal from being flattened by the press. Because there is an extreme concentration of pressure to selected parts of the matrix die when it is used in this way, it is necessary to face the matrix with steel. The two dies also must be accurately aligned. You must provide registration pins for this.

It is advisable to work very accurately and to avoid attempting to push this concept beyond what it will do. Extremely large or small, complex or "pointy" designs, as well as the use of thick metal may cause problems. Hardening the steel could help, but even a small amount of warp or distortion would cause additional problems.

1. Begin by laying out the blanking die as described in the chapter on blanking dies. 15 Then cut pieces the same size from following materials: $\frac{1}{16}$-inch mild steel, brass shim and $\frac{1}{4}$-inch acrylic. Paint one side of the steel and of the acrylic with layout liquid.

15

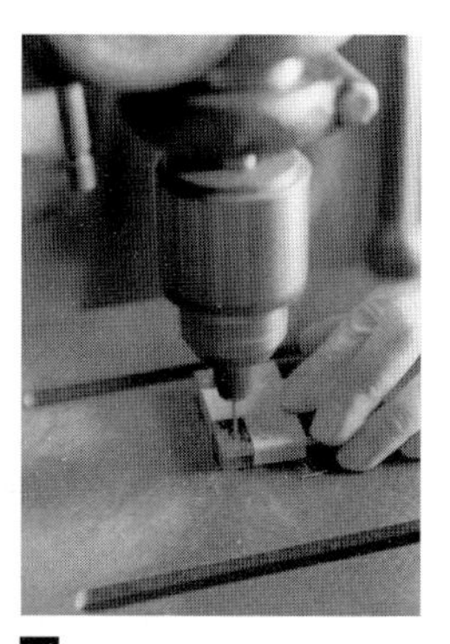

16

2. Next, stack them as follows: 1) acrylic (with the painted side down), 2) 1/16 mild steel (also with the painted side down), 3) the brass shim, and 4) the blanking die on top, face up. Tape them securely together.

3. Three holes are needed for registration pins. [16] Drill them through all layers; two in the upper corners and one below the second set of hinge holes. (If the pins are too close to the cutting edge, they may get in the way later when loading the die for blanking.)

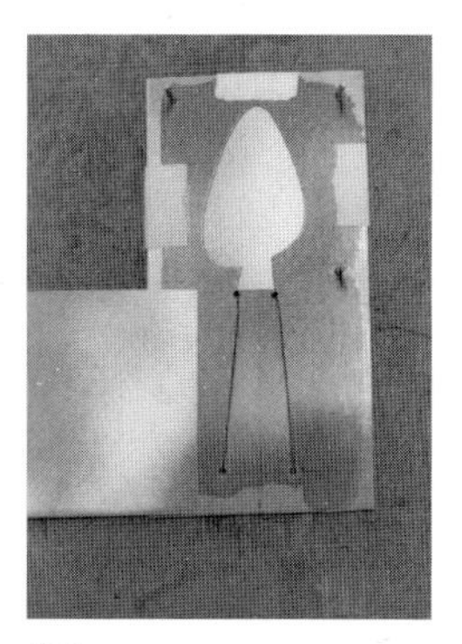

17

4. Remove the tape and disassemble the stack. Then saw out the blanking die, using the appropriate saw blade and angle.

5. Next, cut the shim so that it can be blanked in the die with the registration holes lined up. (The shim must clear the upper set of hinge holes and not cover the third guide pin hole.) [17] Press down on the hinge, place the shim in the die, put guide pins in the registration holes, and tape the shim and die together.

6. Remove the pins and use the press to blank the shim. [18] When you take the die out of the press, turn it over. *The flange taped to the back side of the pancake die is the "up" side of a template for both the acrylic matrix die and its steel face plate.* Note that it is a mirror image of the original shape. This is correct!

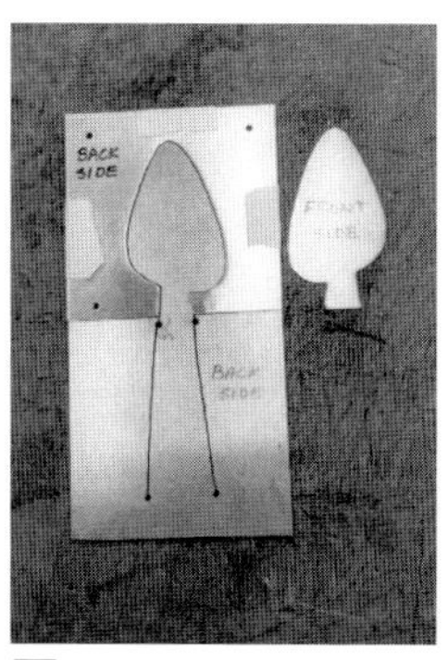

18

7. **19** Use the guide pins to align the shim template when scribing the design onto the coated sides of the matrix die (the acrylic) and on the face plate (mild steel). The original pattern may be used to complete the shape where the hinge is located. Pierce, saw and then file each one up to the line. It is extremely important that this stage be done accurately.

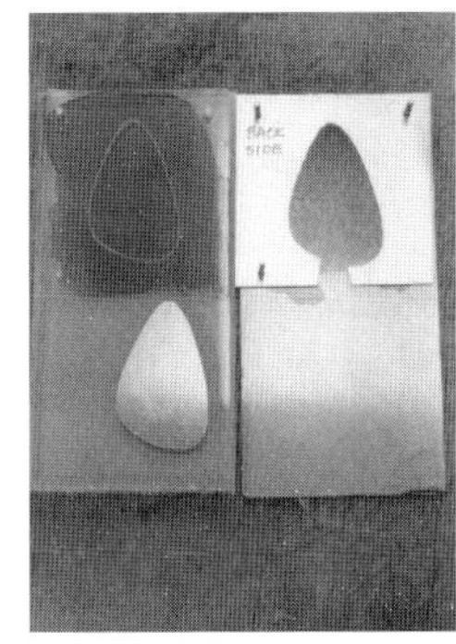

19

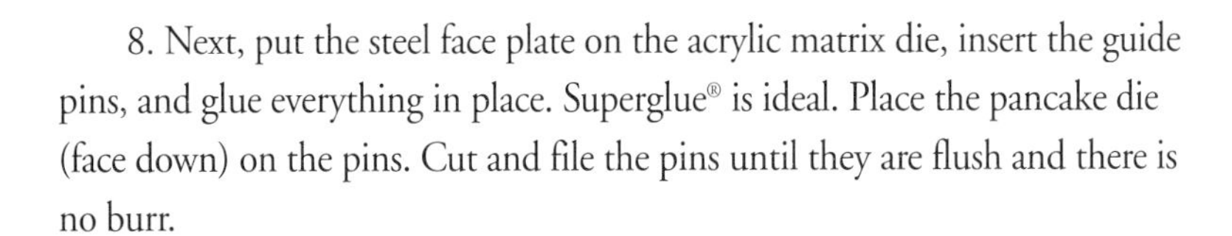

8. Next, put the steel face plate on the acrylic matrix die, insert the guide pins, and glue everything in place. Superglue® is ideal. Place the pancake die (face down) on the pins. Cut and file the pins until they are flush and there is no burr.

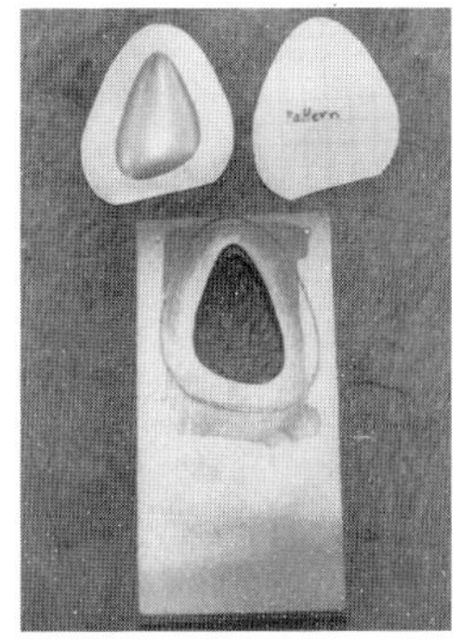

20

9. **20** One more step is to make a pattern (allowing for a ⅜-inch flange) and draw it on the face plate. This is helpful when you cannot see through the matrix die and visually align the metal and the die.

10. Use the matrix die in the usual manner. Several pressings may be necessary to achieve the desired form.

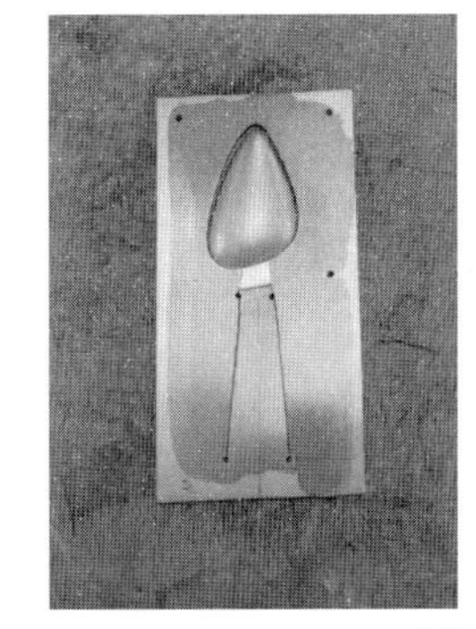

21

11. **21** To blank the formed metal, push the hinge of the blanking die down in the usual manner, and insert the metal. Adjust so that the design is positioned correctly. **22** Turn the die over and place the pancake die onto the matrix die, aligning the registration pins in the holes. Place between two pieces of acrylic and press.

There are many, many ways to combine various die-forming processes with each other and with traditional metalsmithing techniques. This chapter shouldn't be considered the end, but a beginning.

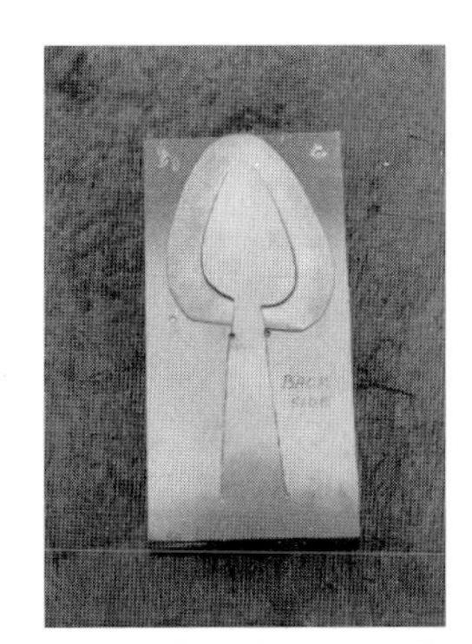

22

g l o s s a r y

ACRYLIC and PLEXIGLASS generic names for polymethyl methacrylate, a cast thermoplastic acrylic resin, sold under the trade names of Lucite® Plexiglas® and Acrylite®.

ANNEALING the process of heating metal to the temperature necessary for recrystallization, and cooling it. Annealing relieves stress and reverses the work hardening process, leaving the metal soft.

ASYMMETRICAL REVERSIBLE MATRIX DIE a silhouette die (or a pair of dies) that has two exactly matched opposite faces. It forms "mirror image" parts which can be fit together to make a hollow asymmetrical form.

BLANK a "cutout" metal shape produced by a die, or the act of producing the cutout.

BLANKING DIE a die that cuts out blanks, such as a pancake die.

BRAKE ASSEMBLY an accessory for the press that holds flat dies for bending and forming metal.

BENDING DIE a flat metal die, used with the brake holder and with urethane pads for bending metal in various ways

BENDING BRAKE a bending die for making 90-degree angles.

CONFORMING DIE a two-part die made of rigid material. Metal is formed by pressing it between the "male" and "female" parts.

CONTAINED URETHANE Urethane contained in a steel cylinder with a welded bottom, used with daps, mushrooms and punches for forming and embossing with maximum efficiency.

DEVCON® PLASTIC STEEL® LIQUID TYPE B a steel-filled epoxy compound consisting of 80% steel and 20% epoxy which can be used to make conforming dies.

DIE any one of the various tools or devices for forming, shaping, stamping or cutting metal.

DIE SETS a matched set of positive and negative steel dies for stamping and/or coining.

DIE SHOE a device used in presses for positioning steel die sets.

DOUBLE-SIDED TAPE double-sided Scotch© tape, available at stationery stores has various uses in die forming.

DRIFT PUNCH a tool that is handy for replacing the rubber stopper from a jack.

DUROMETER a measurement indicating hardness, the higher the number the harder the material.

EMBOSSING DIE a die used for forming a shallow relief in sheet metal.

FACE PLATE a pierced sheet of heavy gauge metal that is placed over a matrix die to ensure accurate and crisp edges.

FINGER BRAKE a bending die for making 90-degree angles in lengths from ½ inch to 5 ¾ inches, used primarily for box making.

FLANGE the extra metal that skirts a shape after it has been formed or blanked.

FLANGELESS DIE a conforming die that forms over its entire surface without leaving a flat flange. Flangeless dies can be used to form precut blanks.

FLEXANE® a two-part pourable urethane. Highly toxic chemicals are released during the mixing and curing of this material. Air purifying respirators do not provide adequate protection and mixing it is not recommended. After the curing period, Flexane® may be used safely in the manner described in this book. Do not heat, grind burn, saw or sand any form of urethane as dangerous fumes are released.

HEAT TREATMENT the process of making high-carbon steel, tool steel, a precise degree of hardness, including two steps, hardening and tempering.

HYDRAULICS the scientific principles that provide the mechanical advantage, and the force supplied by the jack. (The pressure exerted by a small piston is communicated through a confined liquid to a large piston; the force being multiplied as many times as the area of the smaller piston must be multiplied to equal that of the larger.)

INTAGLIO DIE a die in which a design is carved into a surface, and into which metal can be pressed, an embossing die.

KERF a cut made by a saw.

LAYOUT LIQUID a type of stain that is applied to metal before scribing a pattern in order to reduce glare and make the lines more visible.

MARRIED METALS a technique in which different metals are joined by soldering side by side.

MATRIX the negative, or impression part of a die, also called the "female" die.

MILD STEEL any low carbon steel. It cannot be tempered.

MOKUMÉ a traditional technique, sometimes called Japanese woodgrain, made by laminating mixed metals.

MONO-TOOTH SAWBLADE a sawblade with a single spiral "tooth" for cutting metal or soft material.

MULTIPLE MATRIX DIE a matrix die with several cutouts than can form several shapes at once.

NON-CONFORMING DIE a die with only one part, either a matrix or punch, which is used with urethane pads or contained blocks.

OUTRIGGER a metal "blank" used to balance the load when a standoff kit is used.

PANCAKE DIE a simple blanking die made with one cut in a single sheet of tool steel.

PLANISH a process of smoothing and evening out the surface of metal with light overlapping strokes of a hammer.

PLEXIGLASS and ACRYLIC generic names for polymethyl methacrylate, a cast thermoplastic acrylic resin. It is sold under the trade names of Lucite®, Plexiglas® and Acrylite®.

POSITIONER steel plate with a half round groove, used with the brake holder.

PSI a measurement of pressure, pounds per square inch.

PUNCH a tool that is driven against and/or into a surface, sometimes called a "male" die. It is generally used with contained urethane.

QUENCH to suddenly and quickly cool heated metal.

RETAINING RINGS short sections of steel pipe (⅛-inch wall minimum) that contain the steel epoxy used in conforming dies.

REVERSIBLE DIE a die having a symmetrical outline. Parts formed in a symmetrical die can be used back to back to make a hollow form.

REPOUSSÉ or REPOUSSAGE an ancient technique in which metal is formed into relief by working from both the front and back with small tools.

ROLL PRINTING a method of imparting a textured surface to one side of a sheet of metal by passing it through a rolling mill with a template.

SHEARING DIE a die that cuts a shape from a sheet of metal, a blanking or pancake die.

SILHOUETTE DIE a simple matrix die consisting of a thick plate (of plexiglass or other material) with the silhouette of a form cut out, used in the press with urethane pads.

SPACER a block of rigid, non-compressible material (such as steel or acrylic) that is used to take up vertical space in the press so that the hydraulic ram is not fully extended.

SPLIT MATRIX DIE a pair of matrix dies that can be arranged in different ways to achieve different forms.

TEMPERING the process of reducing hardened steel's brittleness by gently reheating to a predetermined temperature.

TOOL STEEL steel containing at least 0.2% carbon, which makes it temperable.

TUBING FORMER a type of bending die for making tubing, consisting of a positioner and a set of steel rods. It is used with the brake holder and urethane pads.

URETHANE a tough, rubberlike material that becomes, under pressure, the other half of a non-conforming die.

WORK HARDENED the hard, brittle state metal reaches after it has been worked, caused by compaction of the crystalline structure, and reversed by annealing.

Pressure Requirements: Approximate and Relative Pressure for Various Types of Dies

(and comparative psi readings for different hydraulic gauges)

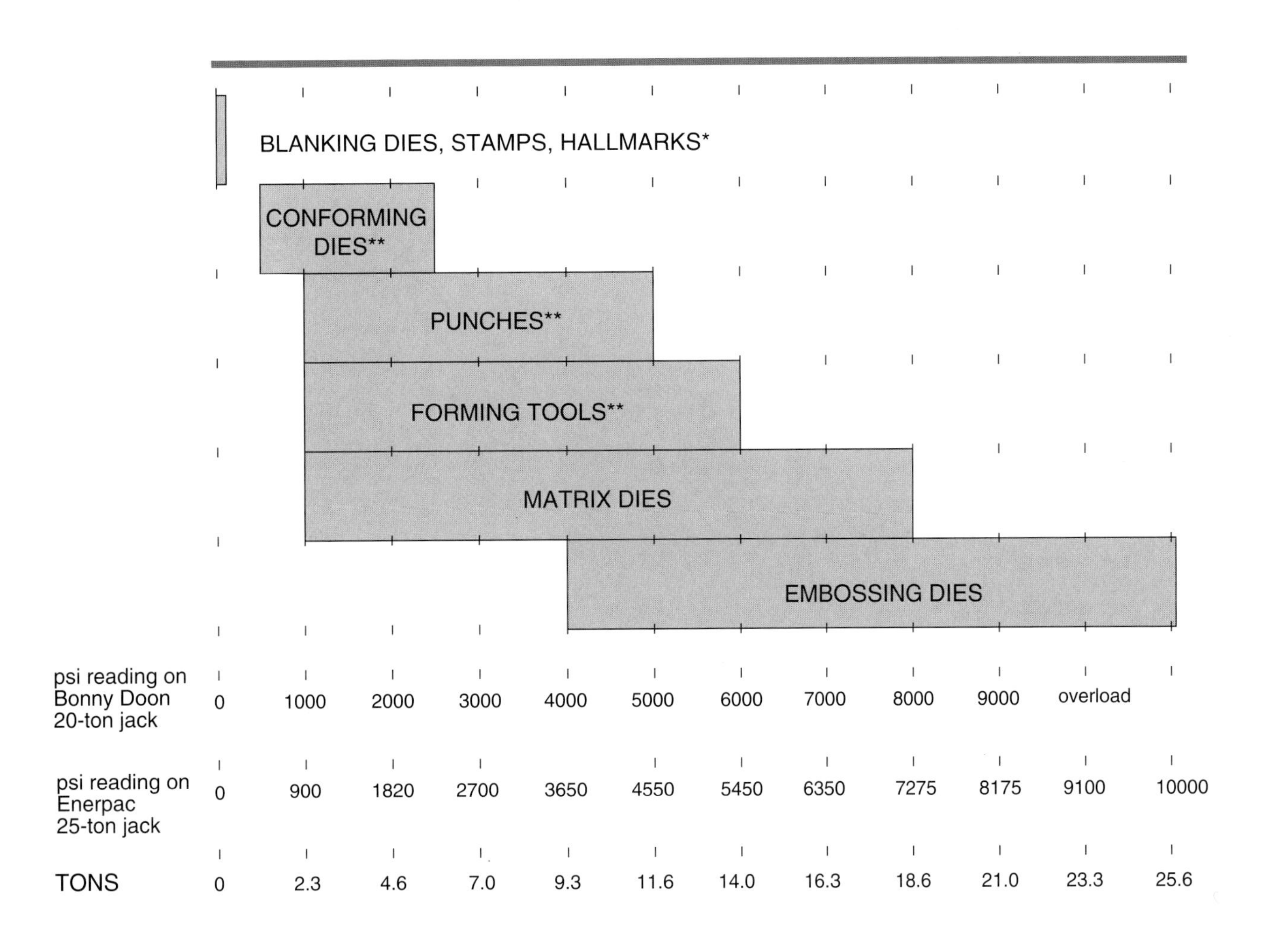

*Blanking dies require very little pressure. Stop when you hear the pop. Stamps and hallmarks require even less. Stop as soon as you begin to feel resistance.

** Conforming dies, forming tools and punches must be monitored visually. Watch what is happening in the press rather than the gauge.

Hole Pattern for Press

(Needed for Bonny Doon Accessories)

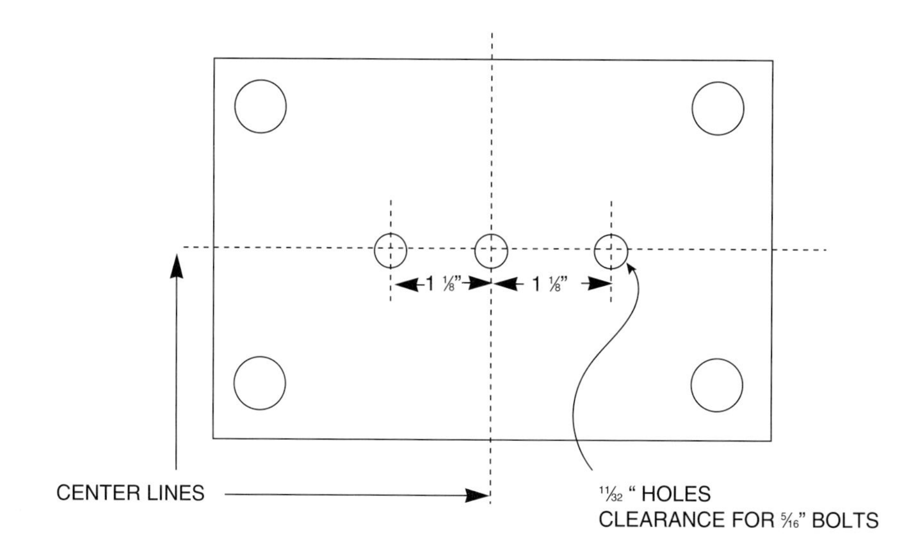

Adjustable Hydraulic Press
Parts and Assembly

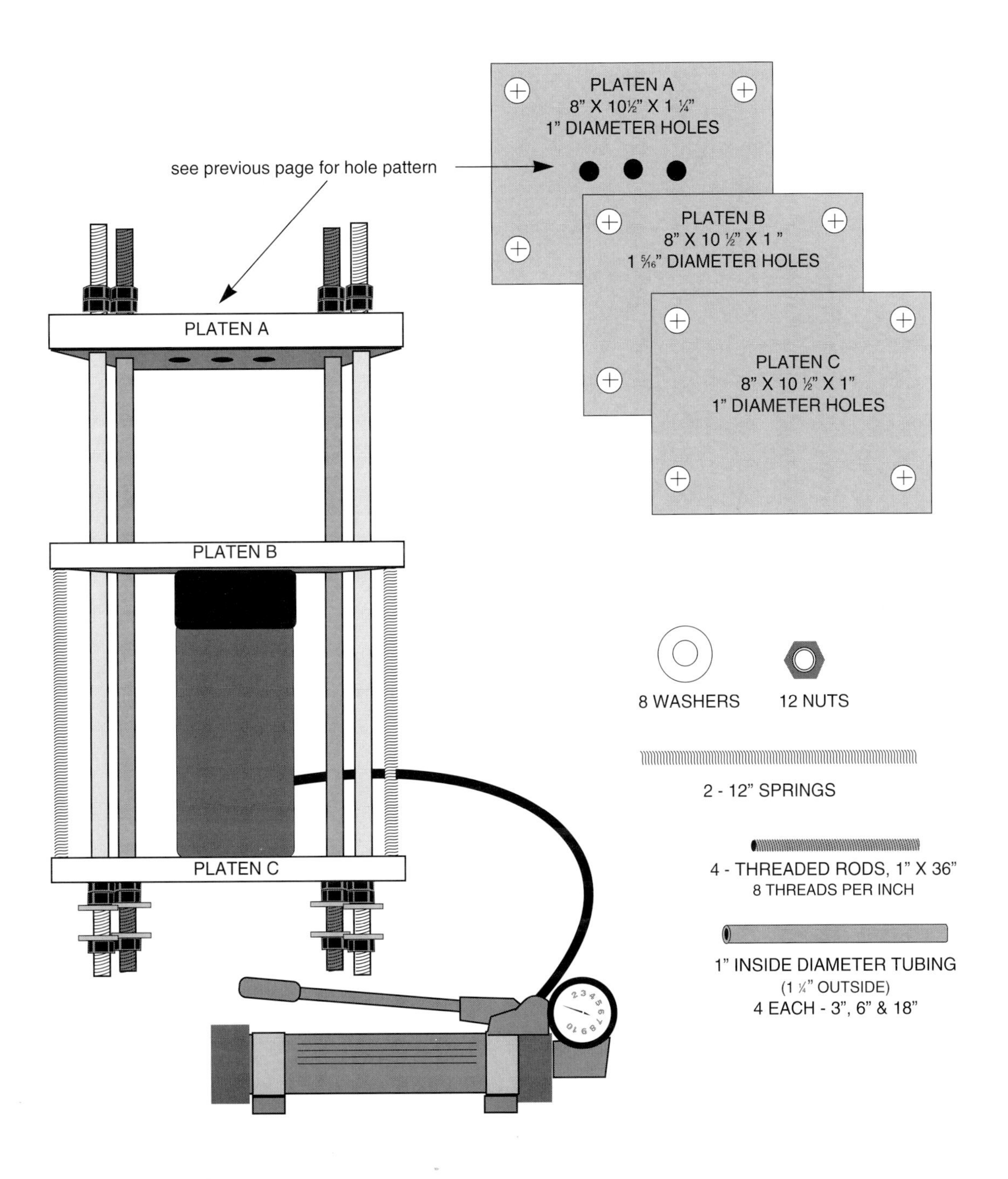

Formula for Blank Width of Large Tubing

Radius = ½ Diameter

[(.446 x thickness of metal) +
Radius (of steel rod)] x 6.283 =
Width of blank

Example:

.81 mm metal thickness
3.15 mm radius

[(.446 x .81) + 3.15)] x 6.283 = 22.06 mm

Example:

.020 inch metal thickness
.125 inch radius

[(.446 x .020) + .125)] x 6.283 = .841 inches

For tubing larger than one inch in diameter, use the same formula and substitute .5 for .446.

Table of Blank Widths for Large Tubing
Measured in Millimeters

METAL THICKNESS	INSIDE DIAMETER									
Inches	¼	5/16	⅜	7/16	½	9/16	⅝	¾	⅞	1
(mm)	6.35	7.94	9.53	11.11	12.70	14.29	15.88	19.05	22.26	25.40
24 G .51 mm	21.35	26.38	31.34	36.37	41.33	46.29	51.32	61.35	71.24	81.23
22 G .64 mm	21.72	26.74	31.71	36.74	41.70	46.66	51.69	61.68	71.61	81.60
20 G .81 mm	22.19	27.22	32.19	37.21	42.18	47.14	52.17	62.16	72.02	82.07
18 G 1.02 mm	22.79	27.81	32.78	37.80	42.77	47.73	52.76	62.75	72.67	82.66
16 G 1.29 mm	23.54	28.56	33.52	38.55	43.51	48.48	53.50	63.49	73.42	83.41
14 G 1.63 mm	24.49	29.51	34.47	39.50	44.46	49.43	54.45	64.44	74.37	84.36

Formulas for Volumes and Quantities of Plastic Steel® for Conforming Dies

To determine the volume of material needed, use the formula for finding the volume of a cylinder:

$$\pi r^2 \text{ x Height} = \text{Volume}$$

The weight that is needed is calculated as below:

1 pound Plastic Steel® = 13.1 cubic inches
1 pound = 453.6 grams
1 cubic inch = 34.63 grams Plastic Steel®

Multiply the volume by 4.63 to find the weight of epoxy that is required.

Volume x 34.63 = Epoxy Weight in grams

If the die is the first half of the conforming die and contains the model, you may subtract the estimated volume that it occupies in the die. If the die is the second half of a conforming die, you need to fill the "female" part of the other die as well as the complete volume of the cylinder, and should estimate and add it on.

The mixing ratio for Plastic Steel® is 90% resin and 10% hardener.

.90 x Epoxy Weight = Weight for resin
.10 x Epoxy Weight = Weight for hardener

Example: A die is 2 inches in diameter and 3/4 inches deep, and contains a small model.

3.14 x (1 x 1) x .75 = 2.36 cubic inches
2.36 x 34.63 = 81.73
round to 82 grams (for total amount needed)
82 x .9 = 73.8 grams resin
82 x .10 = 8.2 grams hardener

Table of Mixing Proportions for Plastic Steel®
for Conforming Dies (Approximate)

	2-inch diameter die		2½-inch diameter die		3-inch diameter die	
½-inch deep	55 gr.	49.5 resin 5.5 hardener	85 gr.	76.5 resin 8.5 hardener	123 gr.	110.7 resin 12.3 hardener
¾-inch deep	82 gr.	73.8 resin 8.2 hardener	128 gr.	115.2 resin 12.8 hardener	184 gr.	165.6 resin 18.4 hardener
1-inch deep	(not recommended)		170 gr.	153. resin 17.0 resin	245 gr.	220.5 resin 24.5 hardener

Table of Sawing Angles for Blanking Dies*

Tool Steel Thickness	Maximum Blank Thickness **	5/0	4/0	3/0	2/0	1/0 (0)	1
1/32 inch .794 mm	22 G .64 mm	17.5	18.	20.	22.	—	—
3/64 inch 1.19 mm	18 G 1.02 mm	12.	13.	15.	17.	18.	19.
1/16 inch 1.59 mm	16 G 1.29 mm	—	9.5	10.5	11.	12.5	14.
5/64 inch 1.98 mm	14 G 1.63 mm	—	—	9.	9.5	10.	10.5

* These angles are recommended for dies that are sawn "free-hand," and for dies that will not be heat treated. One degree may be added if the die will be cut with a precision sawing guide (or power saw), and/or if the die will be heat treated.

** Blanking dies will blank metal up to their own thickness.

Jeweler's Saw Blades: Thickness and Depth (in thousandths) and Drill Equivalents

Number	Thickness	Depth	Drill
8/0	.0060	.0130	80
7/0	.0065	.0135	80
6/0	.0070	.0140	79
5/0	.0080	.0150	78
4/0	.0085	.0170	77
3/0	.0095	.0190	76
2/0	.0100	.0200	75
1/0 or 0	.0110	.0230	73
1	.0115	.0250	71
2	.0135	.0270	70
3	.0140	.0290	68
4	.0150	.0310	67
5	.0160	.0340	65
6	.0190	.0410	58
7	.0200	.0480	55

Weights and Measures

Troy Weight - a standard system of weights used for gems and precious metals

24 grains = 1 pennyweight
20 pennyweights = 1 ounce
12 ounces = 1 pound
(Pennyweight is abbreviated dwt.)

Avoirdupois Weight - a system of weights used chiefly in North America to weigh nearly everything except gems and precious metals (Abbreviated avdp.)

16 drams = 1 ounce
16 ounces = 1 pound

Gram Weight - a unit of weight used in the metric system (approximately one cubic centimeter of water)

1000 grams = 1 kilogram
1000 kilograms = 1 metric ton

Celsius - a temperature scale that registers the freezing point of water as 0° and the boiling point as 100° under normal atmospheric pressure (abbreviated C.)

Fahrenheit - a temperature scale that registers the freezing point of water as 32° and the boiling point as 212° under normal atmospheric conditions (abbreviated F.)

Freezing point = 32° Fahrenheit = 0° Celsius
Boiling point = 212° Fahrenheit = 100° Celsius

Conversions

Inches	x	25.4	=	millimeters
Inches	x	2.54	=	centimeters
Millimeters	x	.04	=	inches
Square centimeters	x	.155	=	square inches
Square inches	x	6.4515	=	square centimeters
Cubic centimeters	x	.06	=	cubic inches
Cubic inches	x	16.387	=	cubic centimeter
Cubic inches	x	.01639	=	liters
US fluid ounces	x	29.5737	=	cubic centimeters
US fluid ounces	x	1.80469	=	cubic inches
US fluid ounces	x	.02957	=	liters
Liters	x	1.0567	=	US quart
Liters	x	61.023	=	cubic inches
US gallons	x	3.785	=	liters
US gallons	x	231.	=	cubic inches
Pennyweights	x	1.555	=	grams
Pennyweights	x	.05	=	troy ounces
Pennyweights	x	0.004167	=	troy pounds
Pennyweights	x	.05486	=	avoirdupois ounces
Pennyweights	x	.0034286	=	avoirdupois pounds
Grams	x	.643	=	pennyweights
Grams	x	.03215	=	troy ounces
Grams	x	.03527	=	avoirdupois ounces
Avoirdupois ounces	x	28.3495	=	grams
Avoirdupois ounces	x	.91146	=	troy ounces
Avoirdupois ounces	x	18.2291	=	pennyweight
Avoirdupois pounds	x	291.667	=	pennyweights
Avoirdupois pounds	x	453.6	=	grams
Avoirdupois pounds	x	14.5833	=	troy ounces
Avoirdupois pounds	x	1.2153	=	troy pounds
Avoirdupois pounds	x	.0625	=	avoirdupois ounces
Avoirdupois pounds	x	.4536	=	kilograms
Kilograms	x	2.2046	=	avoirdupois pounds
Kilograms per square centimeter	x	14.2231	=	avoirdupois pounds per square inch
Avoirdupois pounds per square inch	x	.0703	=	kilograms per square centimeter
Troy ounces	x	31.1035	=	grams
Troy ounces	x	1.0971	=	avoirdupois ounces
Troy ounces	x	20	=	pennyweight
Troy ounces	x	.06857	=	avoirdupois pound
Troy ounces	x	.0833	=	troy pound
Troy pound	x	.82286	=	avoirdupois pounds
Troy pound	x	13.166	=	avoirdupois ounces
Troy pound	x	373.24	=	grams
Celsius	x	(1.8)+32	=	Fahrenheit
(Fahrenheit-32)	x	(.555)	=	Celsius

Comparative Measurements:
B&S gauge, millimeters, thousandths of an inch, fractions/drill sizes

B&S Gauge	mm	inches	drill/ fractions
30	.25	.010	—
29	.29	.011	—
28	.32	.013	—
27	.36	.014	80
—	.37	.015	79
—	.40	.016	1/6
26	.40	.016	78
25	.45	.018	77
24	.51	.020	76
—	.53	.021	75
23	.57	.023	74
—	.61	.024	73
22	.64	.025	72
—	.66	.026	71
21	.72	.028	70
—	.74	.029	69
—	.79	.031	68
—	.79	.031	1/32
20	.81	.032	67
—	.84	.033	66
—	.89	.035	65
19	.91	.036	64
—	.94	.037	63
—	.97	.038	62
—	.99	.039	61
18	1.02	.040	60
—	1.04	.041	59
—	1.07	.042	58
—	1.09	.043	57
17	1.15	.045	—
—	1.18	.047	56
—	1.19	.047	3/64
16	1.29	.051	—
—	1.32	.052	55
—	1.40	.055	54
15	1.45	.057	—
—	1.51	.060	53
—	1.59	.063	1/16
14	1.63	.064	52
—	1.70	.067	51
—	1.78	.070	50
13	1.83	.072	—
—	1.85	.073	49
—	1.93	.076	48
—	1.98	.078	5/64
—	1.99	.079	47
12	2.05	.081	46
—	2.08	.082	45
—	2.18	.086	44
—	2.26	.089	43
11	2.30	.091	—
—	2.37	.094	42
—	2.38	.094	3/32
—	2.44	.096	41
—	2.49	.098	40
—	2.53	.100	39
10	2.59	.102	38
—	2.64	.104	37
—	2.71	.107	36
—	2.78	.109	7/64
—	2.79	.110	35
—	2.82	.111	34
—	2.87	.113	33
9	2.91	.114	—
—	2.95	.116	32
—	3.05	.120	31
—	3.18	.125	1/8
8	3.26	.128	30
—	3.45	.136	29
—	3.57	.141	28
—	3.57	.141	9/64
7	3.66	.144	27
—	3.73	.147	26
—	3.80	.150	25
—	3.86	.152	24
—	3.91	.154	23
—	3.97	.156	5/32
—	3.99	.157	22
—	4.04	.159	21
—	4.09	.161	20
6	4.12	.162	—
—	4.22	.166	19
—	4.31	.170	18
—	4.37	.172	11/64
—	4.39	.173	17
—	4.50	.177	16
—	4.57	.180	15
5	4.62	.182	14
—	4.70	.185	13
—	4.76	.188	3/16
—	4.80	.189	12
—	4.85	.191	11
—	4.91	.194	10
—	4.98	.196	9
—	5.05	.199	8
—	5.11	.201	7
—	5.16	.203	13/64
4	5.19	.204	6
—	5.22	.206	5
—	5.31	.209	4
—	5.41	.213	3
—	5.56	.219	7/32
—	5.61	.221	2
—	5.79	.228	1
3	5.83	.229	—
—	5.94	.234	A
—	5.95	.234	15/64
—	6.05	.238	B
—	6.15	.242	C
—	6.25	.246	D
—	6.35	.250	1/4
—	6.35	.250	E
2	6.54	.258	F
—	6.63	.261	G
—	6.75	.266	17/64
—	6.76	.266	H
—	6.91	.272	I
—	7.04	.277	J
—	7.14	.281	K
—	7.14	.281	9/32
1	7.35	.289	—
—	7.37	.290	L
—	7.49	.295	M
—	7.54	.297	19/64
—	7.67	.302	N
—	7.94	.313	5/16
—	8.03	.316	O
—	8.20	.323	P
0	8.25	.325	—
—	8.33	.328	21/64

Suppliers: Materials and Equipment

DIE FORMING EQUIPMENT AND SUPPLIES

Allcraft Tool and Supply
135 West 29th Street, #402
New York, NY 10001
(800) 645-7124

Bonny Doon Engineering, Inc.
250 Tassett Court
Santa Cruz, CA 95060
(800) 995-9962
www.bonnnydooneng neering.com

Frei and Borel
126 Second Street
Oakland, CA 94604
(800) 772-3456
www.ofrei.com

Rio Grande®
7500 Bluewater Road NW
Albuquerque, NM 87121
(800) 545-6566
www.riogrande.com

"C" FRAME TOOLS for blanking circles and other shapes

Unitool Punch & Die Company
20 Norris Street
Buffalo, NY 14207
(716) 873-8453

CUSTOM-MADE STAMPING TOOLS

Harper Manufacturing Company
3050 Westwood Drive, #B-5
Las Vegas NV 89109
(800) 776-8407

CUSTOM-MADE (PANCAKE-TYPE) BLANKING DIES

Sheltech
4207 Lead SE
Albuquerque, NM 87108
(505) 256-7073
www.sheltech.net

DEVCON® PLASTIC STEEL®
Liquid Type B

Devcon Corporation
(800) 289-4787
www.devcon.com

EPOXY THINNER #PA5022

Anchor Seal, Inc.
20 Riverside Drive
Danvers, MA 01923
(508) 774-5217
www.anchorseal.com

URETHANE

CAM Specialty Products, Inc.
1434 Rayford Road
Spring, TX 77386
(800) 305-3623

TOOLS, TOOL STEEL, LAYOUT LIQUID, DEVCON PLASTIC STEEL

Airgas Direct Industrial
(Formerly Rutland Tool and Supply)
(800) 289-4787
www.rutlandtool.com

HYDRAULIC EQUIPMENT

IMPORTANT: The following hydraulic systems may be used in the adjustable press or in a Bonny Doon frame. Be aware that "specs" and part numbers change, and that the following options are only a few those available. Consult with your nearest hydraulic store/service center for advice and assistance in selecting and purchasing hydraulics to fit your specific needs. ***Professional assembly of hydraulic systems and components is highly recommended.***

BOTTLE JACK WITH GAUGE

Norco
(800) 347-2232
www.norcoindustries.com

Otto Service
2014 Burbank Boulevard
Burbank, CA 91506
(818) 845-3928
www.hyjacks.com

(not pictured)
20-Ton Bottle Jack, Norco KYB #76520G
Gauge #78021

HAND PUMP AND CYLINDER

Enerpac®
www.enerpac.com
800-433-2766

* (Pictured on page 16)
Pump and Cylinder Set #SCR-256H
(hose, fittings and gauge included)

Power Team (a division of SPX Corporation)
(800) 541-1418
www.powerteam.com

*(not pictured)
Pump and Cylinder Set #RPS-256
Adapter "T" #9670
Gauge #905

ELECTRIC PUMP AND CYLINDER

Power Team
* (Pictured on page 19)
"Quarter Horse" Pump # PE 102
Ram #C256C
Pressure regulator #9560
Hose coupler #9798/fitting #9670
Gauge #9051/ Hose

** These cylinders have 25-ton capacities. They are each close to 3.5 inches in diameter and 10.75 inches high. The gauges are 4 inches in diameter and measure accurately in small increments. To use a hydraulic cylinder in a Bonny Doon frame,* **it is critical that the cylinder be fully supported, level, centered and stable.** *A 4-inch section of 2 x 2-inch steel tubing and/or steel shims could be positioned within the base or a bridge/platform could be devised of heavy-duty 4 x 4-inch steel plate.*

TOOLS, METALS AND SUPPLIES

Allcraft Tool and Supply
35 West 29th Street, #402
New York, NY 10001
(800) 645-7124

C. R. Hill Company
2734 W. 11 Mile Road
Berkley, MI 48072
(800) 521-1221

E. B. Fitler & Company
RD 2, Box 176-B
Milton, DE 19968
(800) 346-2497
www.fitler.com

Frei and Borel
126 Second Street
Oakland, CA 9460
(800) 772-3456
www.ofrei.com

Indian Jewelers Supply Company
P.O. Box 1774
Gallup, NM 87305
(800) 545-6540
www.ijsinc.com

Metalliferous, Inc.
34 West 46th Street
New York, NY 10036
(888) 944-0909

Paul H. Gesswein & Company, Inc.
255 Hancock Avenue
Bridgeport, CT 06605
(800) 243-4466
www.gessweinco.com

Rio Grande
7500 Bluewater Road NW
Albuquerque, NM 87121-1962
(800) 545-6566
www.riogrande.com

Swest, Inc.
11090 N. Stemmons Freeway
Dallas, TX 75229
(800) 527-5057
www.swestinc.com

T. B. Hagstoz & Son, Inc.
709 Sansom St.
Philadelphia, PA 19106
(800) 922-1006

T.S.I.
101 Nickerson Street
P.O. Box 9266
Seattle, SA 98109
(800) 246-9984

PLASTICS

Cadillac Plastics
(800) 274-1000
www.cadillacplastic.com

ACRYLITE®

Cyro Industries
(800) 631-5384
www.cyro.com

PLEXIGLAS®

Atofina Chemicals Inc.
Atoglas Division
(800) 523-7500
www.elfatochem.com/newelf/atoglas/

Jewelry and Metalsmithing: Organizations and Periodicals

ORGANIZATIONS:

Society of NorthAmerican Goldsmiths (SNAG)
Dana Singer,
Executive Director
2275 Amigo Drive
Missoula, MT 59808
www.snagmetalsmith.org

Quarterly magazine:
Metalsmith

American Craft Council
72 Spring Street
New York NY 10012
(212) 274-0630
www.craftcouncil.org

Bimonthly magazine:
American Craft

Society of American Silversmiths
Jeffrey Herman,
Executive Director
P.O. Box 704
Chepachet RI 028
(401) 567-7800
www.silversmithing.com

Artist-Blacksmith Association of North America (ABANA)
LeeAnn Mitchell,
Executive Secretary
P.O. Box 816
Farmington, GA 30638
www.abana.org

Quarterly publications:
The Anvil's Ring and
Hammer's Blow

PERIODICALS:

The Crafts Report
P.O. Box 1992
Wilmington, DE 19899
(800) 777-7098
www.craftsreport.com

JCK
(Jewelers Circular Keystone)
201 King of Prussia Road
Radnor PA 19089
www.jckgroup.com

Lapidary Journal
60 Chestnut Avenue
Devon, PA 19333
(800) 676-4336
www.lapidaryjournal.com

Ornament
P.O. Box 2349
San Marcos, CA 92079
(800) 888-8950

HEALTH AND SAFETY INFORMATION:

ACTS FACTS
The Monthly Newsletter from Arts Crafts and Theater Safety
181 Thompson Street, #23
New York, NY 10012
(212) 777-0062

The Artist's Complete Health and Safety Guide
Monona Rossol, author
Allworth Press, New York
1994
www.allworth.com

Center for Safety in the Arts
artswire.org:70/1/csa/

Jewelry and Metalsmithing Books

The Art of Jewelry Design
Deborah Krupenia
Quarry Books
Rockport, MA 1997

The Art of Jewelry Making: Classic and Original Designs
Alan Revere
Sterling Publishing Co., Inc.
New York, NY 1999

Chasing: Ancient Metalworking Technique With Modern Applications
Marcia Lewis
Lamar Productions
Long Beach, CA 1994

The Complete Metalsmith, An Illustrated Handbook
Tim McCreight
Davis Publications
Worcester, MA 1982 (revised 1991)

Contemporary Jewelry
Philip Morton
Holt, Rinehart and Winston
New York, NY, 1970 (revised 1976)

Contemporary Silver
B.S. Rabinovich and H. Clifford
Rizzoli
New York 2000

Creative Gold and Silversmithing
Sharr Choate
Crown Publishers, Inc.
New York, NY 1970

Design and Creation of Jewelry
Robert von Neuman
Chilton
Radnor, PA 1961 (revised 1972)

Encyclopedia of Jewelry-Making Techniques: A Comprehensive Visual Guide to Traditional and Contemporary Techniques
Jinks McGrath
Running Press
Philadelphia, PA 1995

Form Emphasis for Metalsmiths
Heikki Seppa
Kent State University Press
Kent, OH 1978

Jewelry, 7000 Years
Hugh Tait, Ed.
Harry Abrams
New York, NY 1986

Jewelry of the Ancient World
Jack Ogden
Rizzoli
New York, NY 1982

Jewelry Concepts and Technology
Oppi Untracht
Doubleday
New York, NY 1982

Jewelry: Contemporary Design and Technique
Chuck Evans
Davis Publications
Worcester, MA 1983

Jewelry in Europe and America: new times, new thinking
Ralph Turner
Thames and Hudson, Ltd.
London, England 1996

Jewelry: Fundamentals of Metalsmithing
Tim McCreight
North Light Books
Cincinnati, OH 1997

Jewelry Making Manual
Sylvia Wicks
Brynmorgen Press
Cape Elizabeth, ME 1990

The Making of Tools
Alexander Wygers
Van Nostrand Reinhold
New York, NY 1973

Messengers of Modernism: American Studio Jewelry 1940-1960
Toni Greenbaum, Martin Eidelberg
Montreal Museum of Decorative Arts
in association with Flammarion
New York, NY 1996

Metalsmithing for the Artist-Craftsman
Richard Thomas
Chilton
Radnor, PA 1960

The Metalsmith's Book of Boxes and Lockets
Tim McCreight
Hand Book Press
Madison, WI 1999

Metals Technic: A Collection of Techniques for Metalsmiths
Tim McCreight, ed.
Brynmorgen Press
Cape Elizabeth, ME 1997

Metal Techniques for Craftsmen
Oppi Untracht
Doubleday
Garden City, NY 1968

Metalwork and Enameling
Herbert Maryon
Dover
New York, 1971
(original1912)

The New Jewelry: Trends and Traditions
Peter Dormer and Ralph Turner
Thames and Hudson
London, England 1994

Practical Silversmithing and Jewelry
Keith Smith
Van Nostrand Reinhold
New York, NY 1975

Silversmithing
Rupert Finegold and William Seitz
Chilton
Radnor, PA 1983

Silversmithing and Art Metal for Schools Tradesmen and Craftsmen
Murray Bovin
Bovin
Forrest Hills, NY 1963
(Revised 1979)

Silverwork and Jewellery
Henry Wilson
Pitman Publishing Limited
London, England 1978
(Original 1902)

The Story of Jewelry
J. Anderson Black
William Morrow & Co.
New York, NY 1974

Textile Techniques in Metal: For Jewelers, Textile Artists and Sculptors
Arline Fisch
Lak Books
Ashville, NC 1996

Victorian Jewelry, Unexplored Treasures
Ginny Redington Dawes and Corinne Davidov
Abbeville Press
New York, NY 1991

index